T0182017

MAKE YOUR DATA SPEAK

CREATING ACTIONABLE DATA THROUGH EXCEL FOR NON-TECHNICAL PROFESSIONALS

Alex Kolokolov

Apress®

Make Your Data Speak: Creating Actionable Data through Excel For Non-Technical Professionals

Alex Kolokolov
Batumi, Georgia

ISBN-13 (pbk): 978-1-4842-8941-9   ISBN-13 (electronic): 978-1-4842-8942-6
https://doi.org/10.1007/978-1-4842-8942-6

Managing Director, Apress Media LLC: Welmoed Spahr
Acquisitions Editor: Shiva Ramachandran
Development Editor: James Markham
Coordinating Editor: Jessica Vakili

Distributed to the book trade worldwide by Springer Science+Business Media New York, 1 New York Plaza, New York, NY 10004. Phone 1-800-SPRINGER, fax (201) 348-4505, e-mail orders-ny@springer-sbm.com, or visit www.springeronline.com. Apress Media, LLC is a California LLC and the sole member (owner) is Springer Science + Business Media Finance Inc (SSBM Finance Inc). SSBM Finance Inc is a **Delaware** corporation.

For information on translations, please e-mail booktranslations@springernature.com; for reprint, paperback, or audio rights, please e-mail bookpermissions@springernature.com.

Apress titles may be purchased in bulk for academic, corporate, or promotional use. eBook versions and licenses are also available for most titles. For more information, reference our Print and eBook Bulk Sales web page at http://www.apress.com/bulk-sales.

Printed on acid-free paper

Contents

About the Author

Alex Kolokolov is an expert in the business intelligence systems implementation and the development of information panels for managers – dashboards. He is the CEO and founder of the Institute of Business Intelligence, a consulting company specializing in building corporate reporting systems, and author of the training "Make Your Data Speak."

Over 12 years of practice, Alex has set up hundreds of management reports in production, banks, and trading companies; he's worked with software from Microsoft, Qlik, Tableau, and IBM. Based on this experience, he described universal rules for visualization and working with the requirements of business customers.

About the Technical Reviewer

Naveed Ahmed Janvekar is a Senior Applied Scientist in the United States. He works on solving fraud and abuse problems on ecommerce platforms that impact millions of customers in the United States and other parts of the world using machine learning and deep learning. He has more than eight years of expertise in the machine learning space, which includes classification algorithms, clustering algorithms, graph modeling, BERT, and so on. He has a master's degree in Information Science from the University of Texas at Dallas and graduated top of his class and was awarded as scholar of high distinction and inducted in the prestigious International Honor Society Beta Gamma Sigma. He has a Bachelor of Engineering degree in Electronics and Communication from India. He has worked with influential firms such as Amazon, Fidelity Investments, and KPMG. In his current role, he is researching on identifying novel fraud and abuse vectors on ecommerce platforms and using active learning to improve machine learning model performance.

From the Author

My name is Alex Kolokolov. I have been involved in corporate reporting automation projects since 2009. My company is called the Institute of Business Intelligence and specializes in two areas: (1) the implementation of business intelligence systems and (2) training in working with data and visualization.

Over the past years, I have been able to work with industrial holdings, metallurgical plants, banks, state corporations, and international FMCG companies. They are united by a common problem: there is a lot of data, but little understandable information. Every year, new IT systems and analytical divisions appear, but at the strategic level, companies do not see any benefit from this.

Businesses need data visualization turning a stream of numbers into visual information, which helps to quickly grasp the essence and make an informed decision. It looks like a beautiful presentation, but allows you to click on the desired indicator and find out the details down to the initial data. This is called an interactive dashboard.

It turned out that to get such a dashboard, it is not enough to buy a BI system. All these programs are just constructors in which everything needs to be invented from scratch.

An analogy with oratory is appropriate here. The attention of colleagues at meetings is won by a speaker with a beautiful and understandable presentation, and not with a bunch of bulky tables, even with useful calculations.

I want your reports to be clear and eloquent when the data speak the language of business. This skill will definitely help to clearly and convincingly convey Information to clients, colleagues, or managers. That's why I wrote this book.

Who Is This Book For?

Every report has two audiences. The first is the analysts who collect data and make calculations. The second is the consumers of this information, the managers who make decisions based on the collected data.

This book is suitable for both, but it is written in such a way that everything is clear to business users, even without technical knowledge. I presented the material in simple "human" language – just as I explain in my trainings and MBA programs in business schools.

My goal is to show managers with different professional backgrounds how to "tame" data visualization, set a task to develop a dashboard, and control the quality so that the business gets a decision-making tool.

How to Read the Book

The first five chapters are about step-by-step building a dashboard based on data on wage fund expenses. In fact, it is not so difficult to write 200 pages about it. It's just that in this step-by-step case, I included all the popular mistakes and difficulties that are encountered when working in Excel. And I shared life hacks on how to solve or prevent these problems in a few clicks, thereby saving time and nerves.

I tried to make the practical part a visual cheat sheet, and not just instructions for pressing buttons in Excel. Therefore, the actions are divided into steps with illustrations, and something is designed in the form of short checklists. The QR codes at the end of some parts will direct into related video tutorials so you can get more information. Someone likes to read more, it is more convenient for other people to watch – you got two in one.

The final part of the book is about the rules of visualization. I am not trying to say something new here. Rather, I tried to sort through the answer to the eternal question of beginners about how to choose diagrams.

To do this, I took the most popular visual elements as examples and compared them with data analysis types. Such schemes are called chart-choosers, and many have been invented. I don't like any – they are quite complex and hard to remember. My "invention" in this part of the book was only the "visualization compass": I hope it will help you choose the right diagrams for your task.

Usually, in training materials, theory comes first, and only then practice. I did it differently, focusing on my coaching experience. It shows that a skill is acquired better when you first do it, and then you systematize the acquired knowledge with the help of theory.

This Book Is Not About Excel

No matter how much artificial intelligence, robotics, and other latest technologies are discussed, business decisions are still made by people. Often this has to be done in conditions of lack of information, when there is a lot of data, but there are few useful conclusions from them. And Excel, as a basic tool for most, should be mastered at least to solve this problem.

I'm not a fan of Excel, but it's still used even in large corporations, and you can't ignore this. It just so happened: we all love him, hate him, and cannot do without him.

For me personally, Excel is valuable in that you can test a hypothesis right in the source data file and sketch out a draft of a future dashboard or presentation in half an hour. A handy tool for managers and a foundation of digital skills for top managers.

I would say this: no, this book is not about Excel. All principles, rules, and techniques are relevant for working in any BI system. If you know them, then even your first dashboard in BI will do better than beginners usually do.

Three Stories That Made Me Write This Book

Among my clients, there are dozens of companies and hundreds of their employees. Among my students, there are thousands of people with different backgrounds. All of them come to me with problems and tasks, many of them tell interesting stories.

Over the years, stories about the complexities of working with data and visualization have accumulated so much that they formed the basis of this book. I will tell three such stories. They are about many executives, analysts, and managers. And if the book makes the work of at least one of them easier, I will know that I have invested time and energy in it for good reason.

Story 1. How George from the Ministry Made a Summary Report

George heads the development department in one of the ministries. Usually he does not write reports – subordinates do. But sometimes he has to answer for the quality of the data and their presentation himself.

It was in this situation that George found himself when he had to prepare a report on the status of the implementation of a target program for a deputy minister. To do this, it was necessary to collect data from subordinate organizations and demonstrate in an understandable way the situation as a whole.

Problem: There Is Data, There Is No Clarity

At first glance, it was not difficult – the program consisted of a list of projects, activities, milestones with deadlines, and planned results. But it was necessary to take into account a nuance: projects can have different sources of funding and one project can relate to several targeted programs.

All subordinate organizations sent Excel reports on their projects. Some of them added columns with event statuses, others colored rows by status, third ones wrote about the reasons for deviations in the comments to the cell.

Each organization's report showed that everything was supposedly under control, but the overall progress of the program remained unclear. The task of George was to show

- What projects are not being implemented, lagging behind the planned deadlines
- Where is the problem in financing, where are the sources changing
- Which of the responsible persons is not coping
- Which projects were agreed to be postponed

The desired report did not require a beautiful presentation: a table with reliable data would be enough. But there was a lot of data, and it was not easy to collect them into one whole. A concise summary report was required to provide details on each project.

George already knew that this is called a dashboard – he took one of my courses. He even roughly imagined how such a dashboard should look like. But he had absolutely no idea how to fit all the necessary information into it and structure the data. And it was impossible to entrust the task to someone else, because he had to answer personally to the deputy minister.

How to Show Everything at Once?

After suffering in attempts to reduce heterogeneous tables into a single array, George came to me for a consultation. He looked, frankly, not the best way:

I didn't sleep for two days. There are 4 days until the meeting with the Deputy Minister, and my report is not ready yet. Help me to make a form for collecting data from departments. Everyone sends a report, as it suits him, I've been trying to put them together all night, but nothing comes out!

I don't even need beauty on the dashboard. The main thing is that the data converge with each other and can be filtered according to the characteristics of projects.

We discussed what conclusions he wants to convey, what counter-questions may arise, what filters are needed to show the details. As a result, we drew a sketch of the dashboard and compiled a list of fields for a flat table. We took ten lines from the old report and converted it to a new format. We tested it and realized that we need to add a couple more features to the data collection form.

As a result, George received a template and gave the task to his subordinates to fill out on time clearly according to the specified form. He even managed to analyze and propose to the deputy minister a new scheme for managing the target program. And when, on the last day, two departments sent "the most recent correct data," he just loaded them into a template – nothing had to be redone.

Can't Delegate? You'll Have to Learn How to Work with Data

Managers, and even more so top managers, do not collect and process data every day. For them, analytics is not a profession or a hobby, but a regular necessity. Especially when the data is confidential or requires personal attention. It is impossible to simply entrust this to a subordinate, because you need to draw conclusions yourself.

That's why I started this book with the part about data preparation. You don't have to keep all the Excel functions in mind. When such a task arises, you will open this visual guide, remember the principle of operation, and find which button to press. And while others will confer for a week and agree on the format of data presentation, prepare a dashboard in a couple of hours and find insights for strategic decisions.

Story 2. How Analyst Michael Brought the Director the "Aircraft Control Panel"

When we met, Michael had been working for an energy sales company for six years. He said that the management appreciates his technical skills, but... understands very little of everything he does. Mike even liked it: the fact that no one else will be able to do such a job is not bad for self-esteem.

An Hour and a Half to Explain the Report

He prepared weekly reports for the commercial director. The director complained that he did not understand anything, asked millions of questions on "translation from analytical language to human" and hinted that it would be nice to learn how to make beautiful presentations.

"What is more important to you??" Michael was angry. "I spent three days on these calculations, I give you a full calculation, and do you need pictures?!"

"Mike, well, we've been sitting on this for an hour and a half," the director complained. "And so every week. There shouldn't be such a thing that every table has to be explained for so long, do you understand?"

Usually everyone left with their own opinion. The director says that Michael is boring, and Mike says that the director is stupid.

"Make It Clear and Come Back Tomorrow"

The last straw was the factor analysis for the past quarter. There doesn't seem to be anything to reproach Mike with — he worked day and night to make it on time. He came to the meeting exhausted, but he managed to.

It was important to take into account dozens of factors and get a conclusion on what affects the financial result of the company. Here is the amount of productive supply, the dynamics of changes in the amount of released energy, weather factors, and many, many more things.

Wanting to please the commercial director, this time Mikhail decided not to limit himself to numbers, so he added textual explanations of cause-and-effect relationships as well as graphs and diagrams.

As usual, he brought the result to Excel. It turned out to be a huge matrix: tables, diagrams, graphs, buttons for recalculations. It all looked like an airplane control panel for professional pilots.

Mike felt that the commercial director probably wouldn't figure it out. This means that he will have to spend an hour and a half explaining again. But there was no time left for "beauty guidance." The meeting went something like this:

"Look, this time I made a dashboard, as you asked, with visualization!"

"Mike, tomorrow I have to take this to the CEO. What am I going to tell him? He's going to ask questions, and I don't understand anything here. Make it clear and come back in the morning. This is your main goal now, I need a result tomorrow."

Beautiful Is More Expensive

There was no miracle in the morning, the boss still had to delve into all the details. After sitting together for another two hours, they made a short version in the form of a presentation, but it was still impossible to do without large tables. However they learned that the skill of business visualization can be obtained. Then the director sent Michael to my training, where we met.

Analysts often say that all this visualization is only decoration. If we have reliable metrics, then we can draw conclusions even from the table. This is true for the one who collected and calculated this data himself. But for a manager who sees them for the first time, it is important to switch quickly, understand where the problems are, and draw conclusions. And proper visualization accelerates the process of perception, helps to solve this problem.

Like Michael, many analysts are so keen on calculations that there is no time left for design. But this is necessary for business, because it helps to quickly understand the essence and make decisions. And business appreciates it very much.

To simplify this task for techies, I wrote a book in the form of clear instructions with clear checklists. There is no need to break your head and spend long hours on creative searches. Following the best practices, each analyst will be able to make his dashboard look like a designer worked on it.

Story 3. How HR Irene Struggled with Digital Illiteracy

Irene is HR director of a construction company. The head office is located in one of the Manhattan business centers and branches operate in the regions. Irene came to me with something like this problem: everyone has digitized processes in production and sales, dashboards with analytics are set up, and in her directorate "no one needs anything."

HR specialists are located in regional offices with salaries "not above average," and their motivation is the same. They don't know how to work with data, and they don't want to learn.

Not All Employees Are Equally Useful

Subordinates really did not have much interest in work, nor digital literacy. The reports were made without diligence, not for understanding how to increase labor productivity or optimize the budget fund.

Moreover, mistakes were constantly made in the reports. Then they delete a row that seems unnecessary, then they rename the column or add something from themselves. For example, one day in the report card, one person entered the number of employees as "7 in the morning." Of course, this broke not only the numeric format of the cell, but also a bunch of formulas and calculations related to it.

"I Didn't Know You Could Do This in Excel"

We decided to gather all employees in New York and conduct full-time training on working with data. At the training, I saw their faded eyes and unwillingness to change something in their work. But many really just didn't understand what the management wanted from them, why they filled out new tables from month to month.

When everyone saw that their reports were becoming part of the dashboard for the board of directors, the perception changed. Employees began to show interest and offer ideas on what other metrics to add, what conclusions can be drawn from them.

Of course, questions began to arise where to get the data. Part is in ERP, but they are unloaded badly. Part of it is assembled manually, but each branch does it in its own format. I have shown how these tasks can be solved on Excel pivot tables.

As a result, the training participants were divided into two groups. Some, more advanced, started setting up report templates. And others (of which there were more) finally understood how to keep records correctly so that their colleagues could continue to do analytics.

"I didn't know that you could do this in Excel," said an HR specialist from Pittsfield. "I saw similar reports from salesmen, but I thought it was worth millions. And now I can make a beautiful report for the head myself, and also have time to write an explanatory note with conclusions to it."

Digital Talent Pool

A month after the training, I asked Irene how things were going with digitalization. She answered honestly, that many, after leaving their region, found themselves caught in the routine. But, on the other hand, an active group of six people appeared, who systematize reports in their branches and develop new standards.

Now they are not sending her a mountain of tables, but all ready conclusions and suggestions, comments on deviations. And for this, they receive an additional bonus. By the end of the quarter, they plan to replicate the reporting system to other, more passive branches.

Some of the employees did not want to grow, and someone no longer became a performer, but a reliable adviser for their manager. Irene formed a team of like-minded people around her and gave them motivation for career growth.

Of course, this book is not enough to implement the digital transformation of the company. It's not about strategy or artificial intelligence. But if the majority of office employees read it and apply simple rules in their work, then the overall corporate culture will grow seriously. And there will no longer be a gap between advanced managers in Manhattan and "mere mortals" in branches.

If visualization and dashboards are already the standard of work for you, then share this book with your colleagues. And then they will start speaking the same language with you faster, and team efficiency will increase.

Conclusion

These three stories are united by one common situation. The customer has an image of the result, the dashboard itself, but he does not know how to clearly set the task. The performer, the owner of the data is waiting for a clear statement and does not guess what solutions to offer. Because he does not know how it happens differently, from which side to approach the analytical task.

Managers, analysts, middle managers – many people face difficulties in correctly submitting data. I wrote this book so that you can get rid of them as easily and quickly as possible with the help of techniques and practices approved by dozens of my clients and thousands of students.

Approaches to working with data and visualization rules are the same for Excel and for any BI system. I am giving instructions for Excel, because it remains the most common program for working with reports that any analyst and manager has to deal with.

Don't be surprised that the book starts with practice, and theory goes at the end. Everyone has heard the theory one way or another, but stumbled on its application. Therefore, I will take you through the end-to-end case of creating a dashboard – from raw data to advanced design. To do this, I supplemented the book with QR codes and links to video tutorials. They describe the work in even more detail, and the book will be at your fingertips as a synopsis with illustrations.

You will see that all this is not so difficult, sometimes even fascinating, feel your freedom of action and be able to go beyond my checklists by studying the visualization rules in the last part. And also continue to learn useful life hacks.

Remember that you should not draw a diagram just because you know how to do it! It is important to convey the meaning of the data and make the right decision.

Have a good reading and practice at the computer!

Data Preparation

Reports in the usual tables prepared manually are not suitable for building dashboards. The information in them is arranged conveniently and understandable for human perception, but isn't suitable for machine processing. Therefore, dealing with data is the first step before developing a dashboard.

In this part of the book, I will tell you what to pay attention to when analyzing the initial data and show you how to prepare them properly, so that the interactive dashboard is automatically updated.

© Alex Kolokolov 2023
A. Kolokolov, *Make Your Data Speak*,
https://doi.org/10.1007/978-1-4842-8942-6_1

1.1 What's Wrong with the Initial Data

A common problem: employees make reports in a way that is convenient for them without thinking about how others will work with this data later. We get tables that are difficult to understand: with auxiliary calculation columns, comments, and color highlights. Sometimes it is a ready-made report from the Enterprise Resource Planning accounting system, which then "grows" with notes and groupings.

It is convenient for the author of such a table to work with it: to enter data into a new column, drag down the formula, check the calculation results on the next sheet. But the problem is not even that it is not obvious to everyone else — such data cannot be loaded into the pivot tables which are the basis for a dashboard we make.

Let's look at not the worst example – a table with data on staff costs.

Staff costs

Division	January Target	January Actual	February Target	February Actual	March Target	March Actual	April Target	April Actual	May Target	May Actual	June Target	June Actual	TOTAL Target	TOTAL Actual
Total	1,427,671	1,797,655	950,973	867,565	1,066,053	1,166,700	1,766,487	1,875,256	1,357,493	1,224,594	1,251,529	1,229,405	7,820,206	8,161,175
Production Workers	627,980	1,059,594	327,554	340,805	290,608	432,229	638,204	887,568	636,609	484,002	484,002	618,818	3,004,958	3,823,016
Salary	493,547	307,170	231,699	215,687	227,113	347,518	130,200	382,518	465,369	383,634	383,634	465,369	1,931,562	2,101,898
Bonus	0	445,936	0	0	0	0	355,928	307,615	0	0	0	0	355,928	753,551
Civil contracts	10,049	22,455	0	32,661	0	0	0	0	79,573	0	0	24,593	89,622	79,709
Vacation reserve	27,704	74,604	27,259	26,888	21,981	23,444	58,019	39,034	20,093	34,326	34,326	23,444	189,381	221,738
Extra-budgetary funds	96,680	209,429	68,596	65,569	41,514	61,267	94,058	158,401	71,574	66,042	66,042	105,412	438,464	666,120
Commercial Employees	391,330	416,662	291,271	261,818	314,968	332,673	473,888	416,278	374,302	310,861	328,979	258,954	2,174,738	1,997,246
Salary	225,442	145,371	166,044	139,493	189,583	180,151	206,404	163,981	209,738	182,473	182,473	141,886	1,179,684	953,356
Bonus	55,949	133,432	41,356	44,716	38,581	56,908	130,921	137,327	57,473	53,600	53,600	41,603	377,881	467,587
Civil contracts	0	17,198	0	0	0	0	0	0	0	0	0	0	0	17,198
Vacation reserve	24,805	31,245	20,821	21,609	18,182	25,665	34,420	24,693	27,519	6,821	24,939	19,904	150,685	129,937
Extra-budgetary funds	85,134	89,417	63,050	56,000	68,621	69,949	102,142	90,276	79,573	67,968	67,968	55,560	466,488	429,169
Logistics and service	44,475	27,903	10,034	61,916	52,317	53,762	108,274	80,242	49,834	75,946	75,946	33,553	340,881	333,322
Salary	11,917	18,000	900	4,819	30,456	32,267	33,198	37,166	21,270	38,752	38,752	23,904	136,492	154,907
Bonus	6,142	0	0	0	6,142	5,625	42,732	19,191	0	15,268	15,268	0	70,284	40,083
Civil contracts	14,906	1,640	6,879	43,724	0	0	0	0	12,040	0	0	0	33,825	45,364
Vacation reserve	2,243	2,347	117	59	4,638	4,397	9,353	6,821	6,821	5,569	5,569	2,411	28,740	21,604
Extra-budgetary funds	9,268	5,916	2,138	13,314	11,082	11,473	22,992	17,065	9,703	16,357	16,357	7,238	71,539	71,364
Company management	363,886	293,496	322,114	203,025	408,159	348,036	546,121	491,169	296,747	353,784	362,602	318,081	2,299,630	2,007,591
Salary	237,957	108,810	121,350	103,410	186,584	144,103	152,999	134,672	107,850	132,286	132,286	121,748	939,026	745,028
Bonus	38,581	53,600	139,577	57,473	144,103	144,103	227,508	238,985	144,103	184,030	184,030	137,327	877,903	815,518
Civil contracts	0	0	0	0	0	0	0	0	0	0	0	0	0	0
Vacation reserve	11,327	36,815	15,824	10,707	14,874	16,196	44,312	13,308	13,877	6,821	15,638	13,290	115,852	97,135
Extra-budgetary funds	76,021	94,271	45,363	31,437	62,598	43,634	121,303	104,204	30,917	30,648	30,648	45,716	366,849	349,910

- In the rows of the table, there are expense items and divisions, as well as subtotals for them. They are highlighted in color, but technically inside column A, these categories are indistinguishable.

- To the right in the columns, there are the months in the merged cells, and below them is the Target/Actual grouping. That is, we have a two-level "table header."

It is really convenient to enter data into such a matrix structure. Especially if you need to add a number to a cell and forget about it. But we are thinking a few steps ahead: how to set up a report template that will be automatically updated when new data is added.

Solution: The original matrix table needs to be transformed into the so-called "flat" format.

	A	B	C	D	E	F	G	H	I
1	Staff costs								
2									
3	Division		January		February		March		April
4			Target	Actual	Target	Actual	Target	Actual	Target	Actual
5	Total		1,427,671	1,797,655	950,973	867,565	1,066,053	1,166,700	1,766,487	1,875,256
6	Production Workers		627,980	1,059,594	327,554	340,805	290,608	432,229	638,204	887,568
7	Salary		493,547	307,170	231,699	215,687	227,113	347,518	130,200	382,518
8	Bonus		0	445,936	0	0	0	0	355,928	307,615
9	Civil contracts		10,049	22,455	0	32,661	0	0	0	0
10	Vacation reserve		27,704	74,604	27,259	26,888	21,981	23,444	58,019	39,034
11	Extra-budgetary funds		96,680	209,429	68,596	65,569	41,514	61,267	94,058	158,401
12	Commercial Employees		391,330	416,662	291,271	261,818	314,968	332,673	473,888	416,278
13	Salary		225,442	145,371	166,044	139,493	189,583	180,151	206,404	163,981
14	Bonus		55,949	133,432	41,356	44,716	38,581	56,908	130,921	137,327
15	Civil contracts		0	17,198	0	0	0	0	0	0
16	Vacation reserve		24,805	31,245	20,821	21,609	18,182	25,665	34,420	24,693
17	Extra-budgetary funds		85,134	89,417	63,050	56,000	68,621	69,949	102,142	90,276

"Human-readable" format	"Machine-readable" format
Five categories of data in columns and rows:	Each category is in a separate column:
• In the columns – month, target, and actual values • In the rows – the division name and the cost item	1. Division name 2. Staff costs items 3. Month 4. Target values 5. Actual values

If you enter data manually, it will add routine work: for each row with a numeric indicator, you need to repeat the categories (division, article, month). And the compact matrix will turn into hundreds and thousands of rows.

The good news is that if you download the initial data from the database, then any information system allows you to do it in the same "flat format."

How to Convert Data to a Flat Table

A flat table with initial data should contain

- A row of headings with category names
- Five columns by number of categories
- Non-repeating data in rows

Rows shouldn't have empty cells: division, expense item, and month should be specified for each row.

	A	B	C	D	E
1		Staff costs			
2	1	2	3	4	5
3					
4	Division	Expense item	Month	Target	Actual
5	Production Workers	Salary	January	493,547	307,170
6	Production Workers	Bonus	January	0	445,936
7	Production Workers	Civil contracts	January	10,049	22,455
8	Production Workers	Vacation reserve	January	27,704	74,604
9	Production Workers	Extra-budgetary funds	January	96,680	209,429
10	Commercial Employees	Salary	January	225,442	145,371
11	Commercial Employees	Bonus	January	55,949	133,432
12	Commercial Employees	Civil contracts	January	0	17,198
13	Commercial Employees	Vacation reserve	January	24,805	31,245
14	Commercial Employees	Extra-budgetary funds	January	85,134	89,417
15	Logistics and service	Salary	January	11,917	18,000
16	Logistics and service	Bonus	January	6,142	0
17	Logistics and service	Civil contracts	January	14,906	1,640
18	Logistics and service	Vacation reserve	January	2,243	2,347
19	Logistics and service	Extra-budgetary funds	January	9,268	5,916
20	Company management	Salary	January	237,957	108,810
21	Company management	Bonus	January	38,581	53,600
22	Company management	Civil contracts	January	0	0
23	Company management	Vacation reserve	January	11,327	36,815
24	Company management	Extra-budgetary funds	January	76,021	94,271

There are several ways to convert the initial data into a flat table: by manually copying data into the desired cells and by using macros or Power Pivot techniques.

Let's start with manual data copying: this method will take a little time, but for a new user, it's easier than the others.

■ **Tip!**   *If you are a beginner in Excel, then before starting work, make a copy of the worksheet with data so that you can always clarify the original information.*

Step 1. Column A in the initial table contains two categories of data: "Division" and "Expense item." In a flat table, they should be in different columns. Here's how to separate them.

a) Add a new column to the left of column A. Method 1, the simplest: select column A, open the context menu with the right mouse button, select "Insert." Method 2: put the cursor on any cell in column A, in the menu on the "Home" tab, select the "Insert..." button in the "Cells" section and the "Insert Sheet Columns" button in the submenu.

b) Drag the values of cells that contain the names of divisions into a new column. To do this, select the cells, bring the cursor to the border of this block and move it to a new location.

c) Fill down empty values in the new column with the names of divisions for the rows in which the expense items remain.

d) Give to columns A and B the correct names in the row above the data: "Division" and "Expense item," respectively. In the same row, we will specify the headers of the remaining columns.

	A	B	C
4			Target
5	Total		1,427,671
6	Production Workers		627,980
7	Production Workers	Salary	493,547
8	Production Workers	Bonus	0
9	Production Workers	Civil contracts	10,049
10	Production Workers	Vacation reserve	27,704
11	Production Workers	Extra-budgetary funds	96,680
12	Commercial Employees		391,330
13	Commercial Employees	Salary	225,442
14	Commercial Employees	Bonus	55,949
15	Commercial Employees	Civil contracts	0

Step 2. Then you need to delete the rows that contain totals and subtotals by divisions from the table. Otherwise, the data will be summed several times, and we will get an incorrect result.

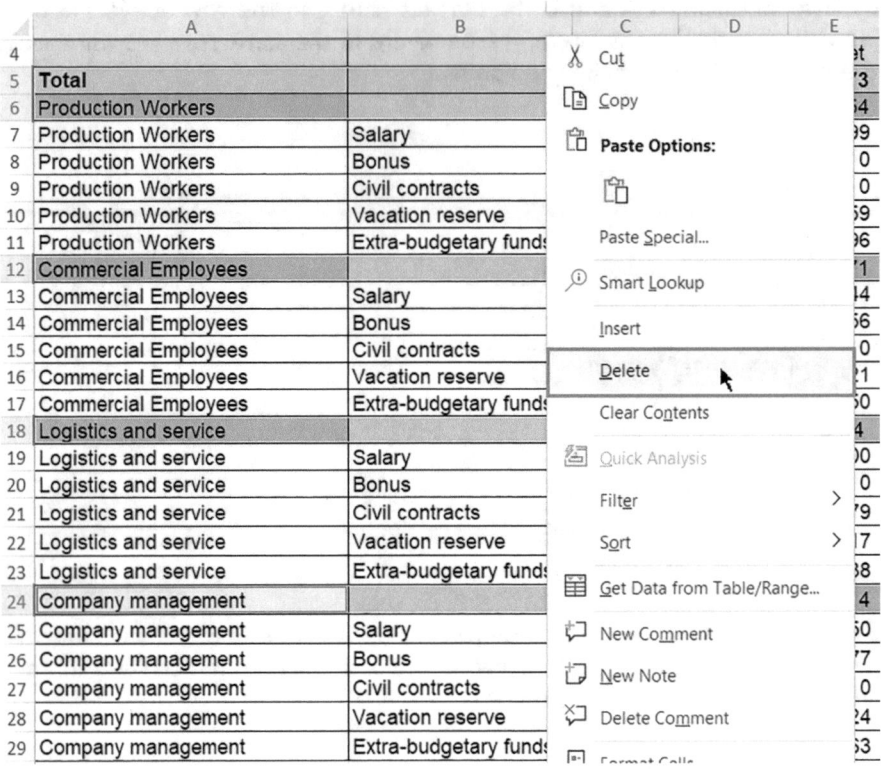

Step 3. The next step is to add a column which will contain data about months.

a) Insert a new column to the left of column C with expense items.

b) Copy the name of the month to the first empty cell of the new column.

c) Select this cell and drag it down by the lower right corner of the frame – the column will automatically be filled with months in their order (the same will happen with a date or a sequence of numbers).

d) Changing this sequence, because first we need to collect data for January. In the lower right corner of the selected block, click the "Autofill Options" menu icon that appears and select "Copy cells."

After completing these steps, we received a flat table in columns A–E for the necessary categories with data for January.

	A	B	C	D	E
1	**Staff costs**				
2		1	2	3	4 5
3					
4	Division	Expense item	Month	Target	Actual
5	Production Workers	Salary	January	493,547	307,170
6	Production Workers	Bonus	January	0	445,936
7	Production Workers	Civil contracts	January	10,049	22,455
8	Production Workers	Vacation reserve	January	27,704	74,604
9	Production Workers	Extra-budgetary funds	January	96,680	209,429
10	Commercial Employees	Salary	January	225,442	145,371
11	Commercial Employees	Bonus	January	55,949	133,432
12	Commercial Employees	Civil contracts	January	0	17,198
13	Commercial Employees	Vacation reserve	January	24,805	31,245
14	Commercial Employees	Extra-budgetary funds	January	85,134	89,417
15	Logistics and service	Salary	January	11,917	18,000
16	Logistics and service	Bonus	January	6,142	0
17	Logistics and service	Civil contracts	January	14,906	1,640
18	Logistics and service	Vacation reserve	January	2,243	2,347
19	Logistics and service	Extra-budgetary funds	January	9,268	5,916
20	Company management	Salary	January	237,957	108,810
21	Company management	Bonus	January	38,581	53,600
22	Company management	Civil contracts	January	0	0
23	Company management	Vacation reserve	January	11,327	36,815
24	Company management	Extra-budgetary funds	January	76,021	94,271

Step 4. Move the target and actual data for February to columns D and E below the January values. Drag down the value "February" opposite these cells in column C.

21	Company management	Bonus	January	38,581	53,600
22	Company management	Civil contracts	January	0	0
23	Company management	Vacation reserve	January	11,327	36,815
24	Company management	Extra-budgetary funds	January	76,021	94,271
25			February	231,699	215,687
26			February	0	0
27			February	0	32,661
28			February	27,259	26,888
29			February	68,596	65,569
30			February	166,044	139,493
31			February	41,356	44,716
32			February	0	0
33			February	20,821	21,609
34			February	63,050	56,000
35			February	900	4,819
36			February	0	0
37			February	6,879	43,724
38			February	117	59
39			February	2,138	13,314
40			February	121,350	103,410
41			February	139,577	57,473
42			February	0	0
43			February	15,824	10,707
44			February	45,363	31,437

Step 5. Repeat step №4 with the data for the remaining months. The names of all the months are moved to column C, the planned indicators – in column D, and the actual ones – in column E.

Step 6. The contents of columns A and B are duplicated as follows by copying or dragging down, thus filling in empty cells.

13	Logistics and service	Bonus	January
14	Logistics and service	Civil contracts	January
15	Logistics and service	Vacation reserve	January
16	Logistics and service	Extra-budgetary funds	January
17	Company management	Salary	January
18	Company management	Bonus	January
19	Company management	Civil contracts	January
20	Company management	Vacation reserve	January
21			January
22			February
23			February
24			February
25			February
26			February
27			February

That's all, it remains to clear the sheet with a flat table: remove unnecessary rows with subtotals and top rows until the category headers.

	A	B	C	D	E
1	Division	Expense item	Month	Target	Actual
2	Production Workers	Salary	January	493,547	307,170
3	Production Workers	Bonus	January	0	445,936
4	Production Workers	Civil contracts	January	10,049	22,455
5	Production Workers	Vacation reserve	January	27,704	74,604
6	Production Workers	Extra-budgetary funds	January	96,680	209,429
7	Commercial Employees	Salary	January	225,442	145,371
8	Commercial Employees	Bonus	January	55,949	133,432
9	Commercial Employees	Civil contracts	January	0	17,198
10	Commercial Employees	Vacation reserve	January	24,805	31,245
11	Commercial Employees	Extra-budgetary funds	January	85,134	89,417
12	Logistics and service	Salary	January	11,917	18,000
13	Logistics and service	Bonus	January	6,142	0
14	Logistics and service	Civil contracts	January	14,906	1,640
15	Logistics and service	Vacation reserve	January	2,243	2,347
16	Logistics and service	Extra-budgetary funds	January	9,268	5,916
17	Company management	Salary	January	237,957	108,810
18	Company management	Bonus	January	38,581	53,600
19	Company management	Civil contracts	January	0	0
20	Company management	Vacation reserve	January	11,327	36,815
21	Company management	Extra-budgetary funds	January	76,021	94,271
22	Production Workers	Salary	February	231,699	215,687
23	Production Workers	Bonus	February	0	0
24	Production Workers	Civil contracts	February	0	32,661
25	Production Workers	Vacation reserve	February	27,259	26,888
26	Production Workers	Extra-budgetary funds	February	68,596	65,569
27	Commercial Employees	Salary	February	166,044	139,493
28	Commercial Employees	Bonus	February	41,356	44,716

How to Simplify the Process of Copying Cells

"Ctrl" is used for Windows. For MacBook, you should use "cmd" instead.

If you select cells, press Ctrl+C (copy), put the cursor in the right place and press Ctrl+V (paste), it will take a long time. There are a couple of ways to speed up this process.

Solution 1. Select the cells, bring the cursor to the border of the selected block and press Ctrl – a "+" appears near the cursor. Hold down the Ctrl key and drag the copy of the data to the desired location with the mouse.

Solution 2. Select the desired cells and double-click on the lower right corner of the selected block: the cells will be filled below.

The result will be the same, but I prefer the second method – it's faster.

Summary

Analysis of the original cross-table showed that it's not suitable for creating an interactive dashboard.

We have identified five categories of data and transformed the table.

1. Distributed five categories of data in five columns

2. Deleted rows with subtotals

3. Filled in the rows with the corresponding data

As a result of these steps, we received a flat table suitable for "machine" processing and ready to create pivot tables – the basis for an interactive dashboard.

■ **Important!**   I'm not encouraging you to move values from columns to rows manually. In our real projects, we have tens of thousands of rows and even millions. It's important to me that you feel the logic of a flat table in our training example "at your fingertips" and could explain to a technical specialist how to properly unload data from the database.

Download the source table

1.2 Preparing the Basis for the Dashboard

The basis of an interactive dashboard in MSs Excel is pivot tables. In this part, you will learn how to create such tables, update data in them, and prepare subsets of data for future visual elements.

Creating a Pivot Table

To create a pivot table, there is no need to select an entire flat table – just put the cursor on any cell and select the "Pivot Table" button on the "Insert" tab.

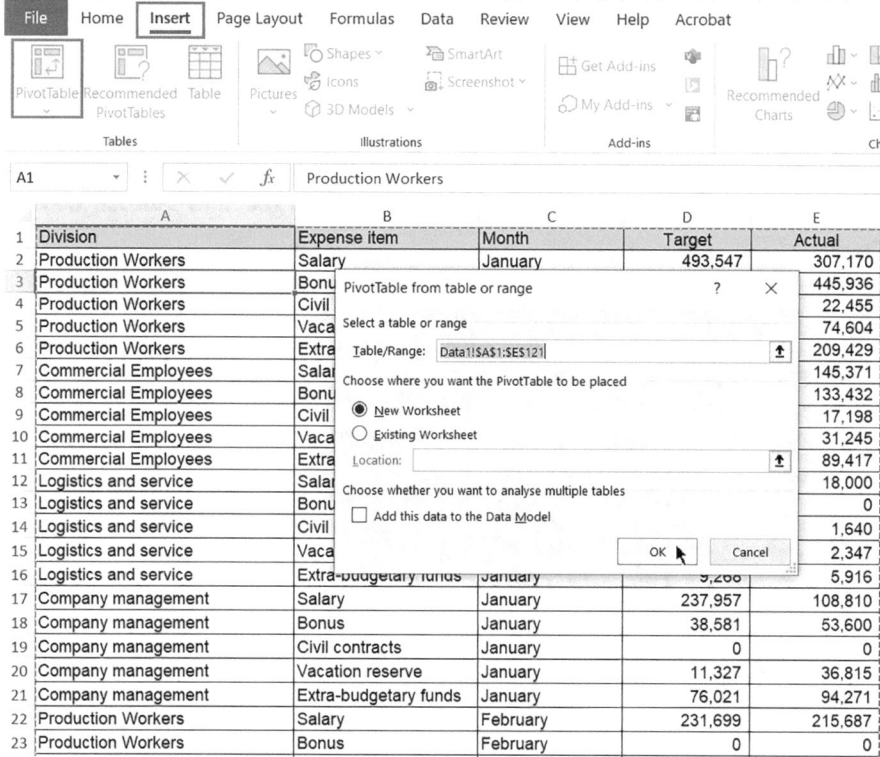

In the appeared dialog window, make sure that the entire required range of data is included. The pivot table must be placed on a new sheet (this is the default option, so you can just click "OK").

On the new sheet, you will have a panel on the right (default view):

- filters

- columns

- rows

- values

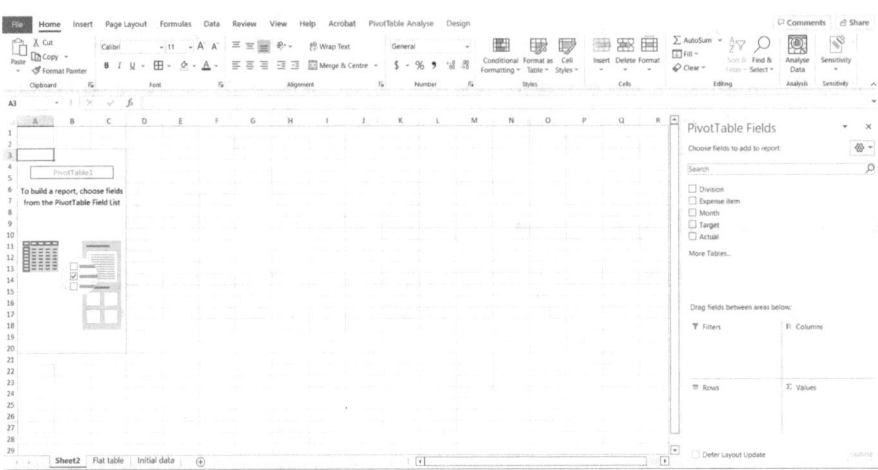

How It Works

Numerical data falls into the "Values" (check the "Target" and "Actual" boxes).
Data categories fall into rows (check the box "Month").

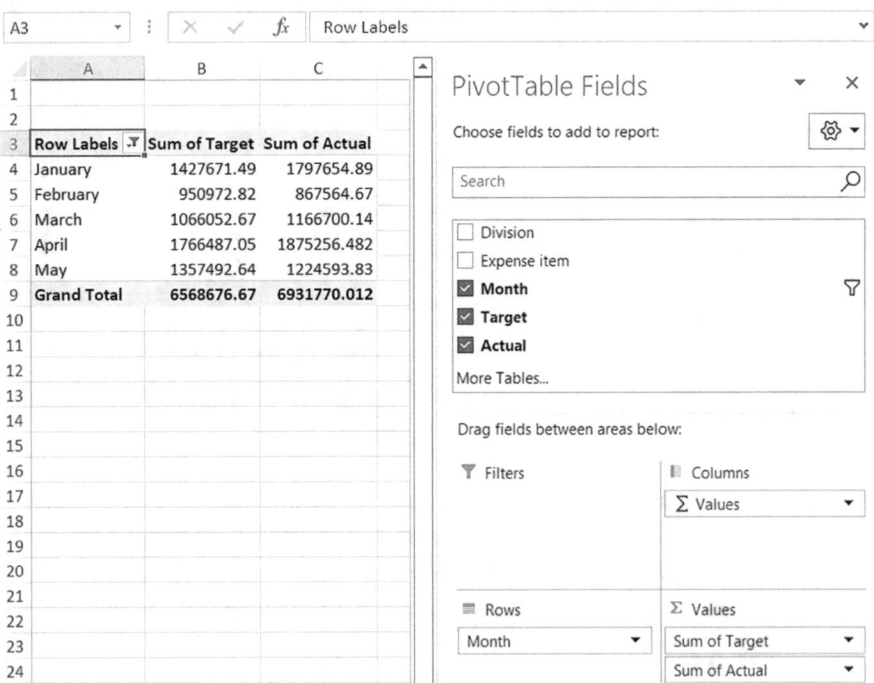

If we add another field with divisions, their names will fall into the rows. They can be moved to columns by dragging and dropping.

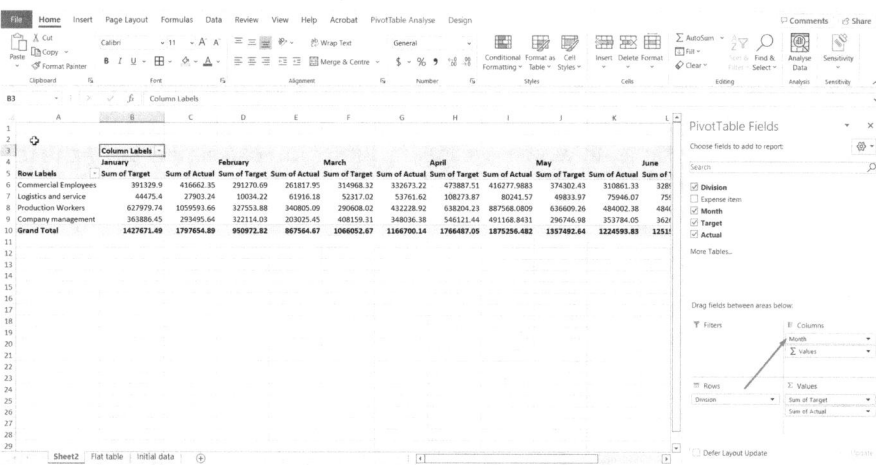

But for building a dashboard, we don't need it – first we make separate simple tables for each graph. If something went wrong, there is a "Clear" button on the "PivotTable Analyse" tab – use it and repeat.

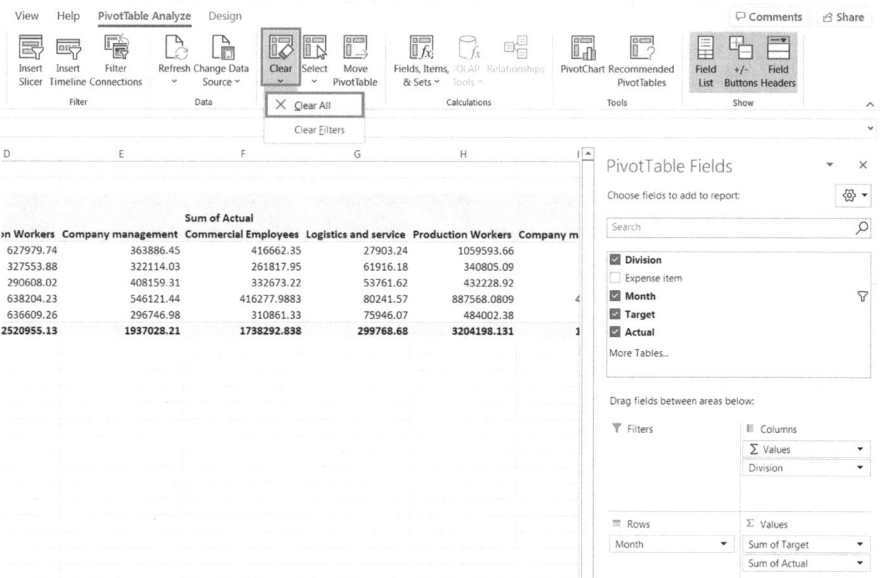

How to Update Data Correctly

The point of a business dashboard is to set up a beautiful output report form once, and then only upload new data. The charts should update automatically. But this is not so easy to do in Excel.

In the previous step, we got a pivot table in which we see the Actual values for staff costs by month from January to May. Now let's do a test: add a row to the original flat table, in which we specify June (you can simply copy the last row of the table and change the month).

	Division	Expense item	Month	Target	Actual
98	Company management	Bonus	May	144,103	184,030
99	Company management	Civil contracts	May	0	0
100	Company management	Vacation reserve	May	13,877	6,821
101	Company management	Extra-budgetary funds	May	30,917	30,648
102	Production Workers	Salary	June	383,634	465,369
103	Production Workers	Bonus	June	0	0
104	Production Workers	Civil contracts	June	0	24,593
105	Production Workers	Vacation reserve	June	34,326	23,444
106	Production Workers	Extra-budgetary funds	June	66,042	105,412
107	Commercial Employees	Salary	June	182,473	141,886
108	Commercial Employees	Bonus	June	53,600	41,603
109	Commercial Employees	Civil contracts	June	0	0
110	Commercial Employees	Vacation reserve	June	24,939	19,904
111	Commercial Employees	Extra-budgetary funds	June	67,968	55,560
112	Logistics and service	Salary	June	38,752	23,904
113	Logistics and service	Bonus	June	15,268	0
114	Logistics and service	Civil contracts	June	0	0
115	Logistics and service	Vacation reserve	June	5,569	2,411
116	Logistics and service	Extra-budgetary funds	June	16,357	7,238
117	Company management	Salary	June	132,286	121,748
118	Company management	Bonus	June	184,030	137,327
119	Company management	Civil contracts	June	0	0
120	Company management	Vacation reserve	June	15,638	13,290
121	Company management	Extra-budgetary funds	June	30,648	45,716

Then go back to the pivot table and so far we don't see June. It seems logical that you need to click the "Refresh all" button on the "Data" tab. But even this does not give results.

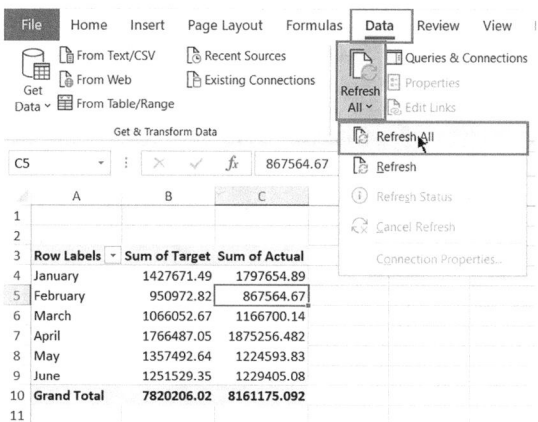

Many skip this step, and then suffer a lot by adding new data to the pivot table. Let's figure out how to set up this process.

Solution 1. Change the Range

When creating a pivot table, Excel highlighted a fixed range of cells with a dotted line. If you change the value inside it, this data will be updated in the report. But our new line with June is outside this range. To add it, go to the menu tab "PivotTable Analyse" and click "Data Source."

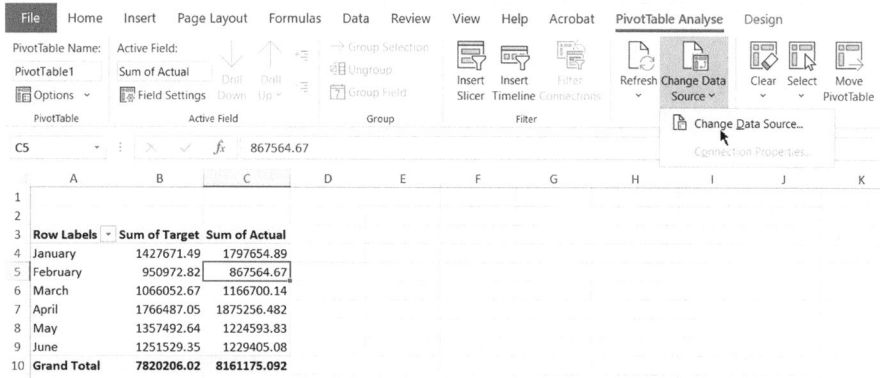

In the dialog box that opens, you need to change the range of cells: to do this, click on the up arrow button and select the cells for analysis on a flat table sheet with the mouse. After that, the data in the pivot table will change.

	A	B	C	D	E
1	Division	Expense item	Month	Target	Actual
2	Production Workers	Sa	Change PivotTable Data Source	? ✕	7,170
3	Production Workers	Bo			5,936
4	Production Workers	Civ	Choose the data that you want to analyse		2,455
5	Production Workers	Va	◉ Select a table or range		4,604
6	Production Workers	Ex	Table/Range: 'Flat table'!$A$1:$E$121 ⬆		9,429
7	Commercial Employees	Sa	○ Use an external data source		5,371
8	Commercial Employees	Bo			3,432
9	Commercial Employees	Civ	Choose Connection...		7,198
10	Commercial Employees	Va	Connection name:		1,245
11	Commercial Employees	Ex			9,417
12	Logistics and service	Sa	OK Cancel		8,000
13	Logistics and service	Bo			0
14	Logistics and service	Civil contracts	January	14,906	1,640
15	Logistics and service	Vacation reserve	January	2,243	2,347

This solution works correctly, but it's the most inconvenient. With any addition of new rows, you'll need to change the range manually. And when there is a lot of data in a flat table, it is easy to make a mistake and not capture some columns or rows.

Solution 2. Select the Entire Columns

I've observed how many Excel users have "automated" data updates. When building the pivot table, they highlighted not the table with data, but all the columns, including the empty rows below. That is, they took the maximum range of rows until the very end of the sheet.

With this method, new rows are added to the pivot table by clicking the "Update All." But the downside is that in each table there will be a row "(empty)."

	A	B	C	D	E
1					
2					
3	Row Labels ▾	Sum of Target	Sum of Actual		
4	January	1427671.49	1797654.89		
5	February	950972.82	867564.67		
6	March	1066052.67	1166700.14		
7	April	1766487.05	1875256.482		
8	May	1357492.64	1224593.83		
9	June	1251529.35	1229405.08		
10	(blank)				
11	Grand Total	7820206.02	8161175.092		
12					

Of course, an empty value can be hidden using filters. It's not that difficult, but in real corporate reports, this "(empty)" constantly pops up on the chart or in the filter and annoys bosses.

In general, this way of configuring the pivot table update is not optimal, it requires additional actions to hide empty values.

Solution 3. Format as a Table

Before inserting a pivot table, let's transform the original flat table into a so-called "smart table."

To do this, select the cell or the range of the table then on the "Home" tab, click the "Format as a table" and select any format from recommended. Formatting in this case is not just a template for formatting cells on a sheet but storing this range as a separate object inside Excel.

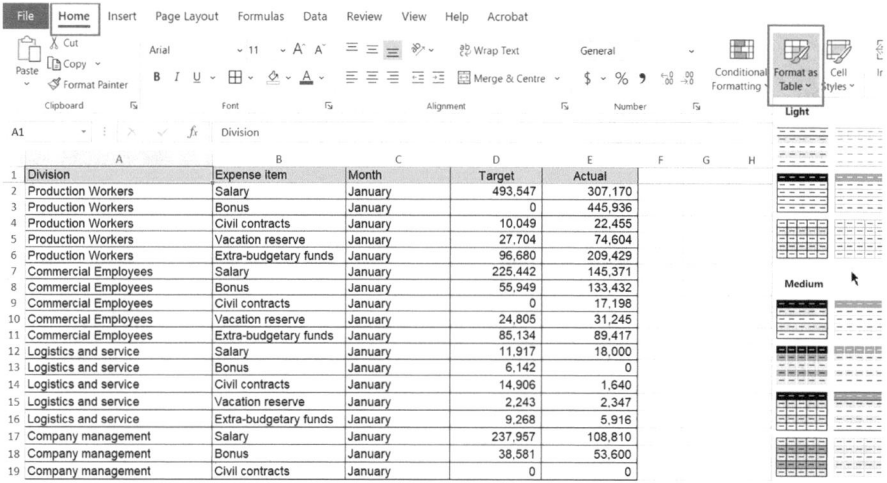

Then, in the window that appears, be sure to check that the "My table has headers" is checked, and click OK.

	Division	Expense item	Month	Target	Actual
1	Division	Expense item	Month	Target	Actual
2	Production Workers	Salary	January	493,547	307,170
3	Production Workers	Bonus		0	445,936
4	Production Workers	Civil contr		10,049	22,455
5	Production Workers	Vacation r		27,704	74,604
6	Production Workers	Extra-budg		96,680	209,429
7	Commercial Employees	Salary		225,442	145,371
8	Commercial Employees	Bonus		55,949	133,432
9	Commercial Employees	Civil contr		0	17,198
10	Commercial Employees	Vacation reserve	January	24,805	31,245

Create Table ? ×

Where is the data for your table?

$A$1:$E$121

☑ My table has headers

OK   Cancel

A smart table always has a filter in the column headers. Additionally, the tab "Table Design" is available for it, where you can set the name of the table (by default it is "Table 1") and immediately create a pivot table based on it by clicking on the "Pivot Table" button.

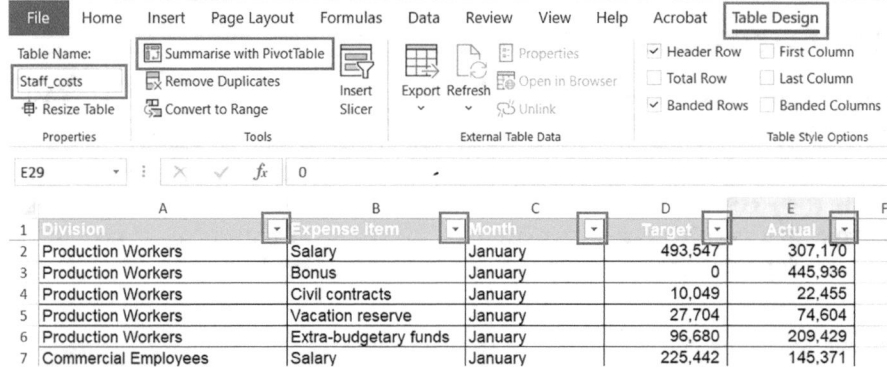

When creating a pivot table, not a range of cells will be specified, but the name of the smart table.

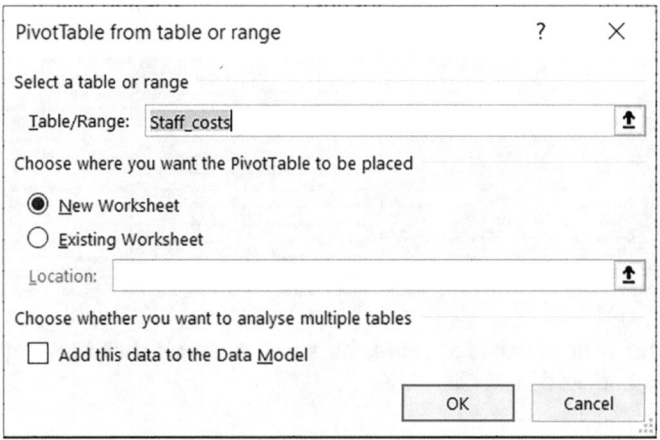

Tip!   *Make it a rule to always give a name to the table: so you'll always know what data you are working with. For example, our smart table can be given a name - "Wage Fund."*

In the future, when adding data to the initial table, Excel will automatically expand the range of the smart table and you'll not be required to do any additional actions, but only click on the "Refresh" button on the "Data" tab. And no empty rows.

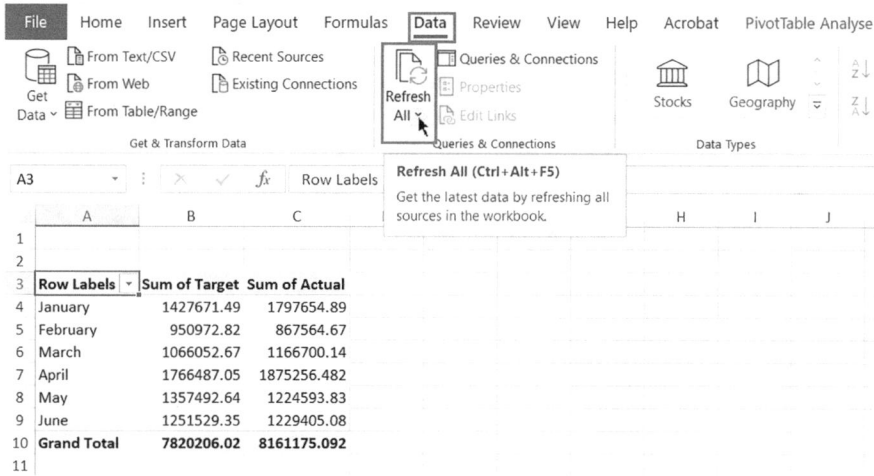

Summary

At the stage of creating pivot tables, it's important to remember that they'll have to be updated – this is the only way the data on the dashboard will remain relevant. But this doesn't happen in Excel on its own.

A life hack that will save you effort – format the prepared flat table as "smart."

- The smart table can be given a name. This is convenient, because when further creating pivot tables, it will always be clear what data they contain.

- This approach saves time. When changing a flat table, you don't have to do additional manipulations: to update the pivot table, clicking the "Refresh All" button will be enough.

1.3 Preparing Data Samples for Visualization

At the heart of each visual element on the dashboard is a separate pivot table report. In general, by the term "report" I understand a complex data representation consisting of several tables and diagrams. Therefore, I will call individual pivot tables data samples.

For the dashboard "Wage fund analysis," we need three samples:

1) The payments dynamics

2) Expenses by division

3) Expenses by item

Let's create them on the prepared data and select the appropriate charts and graphs for every case. We will not consider the issue of choosing correct visual elements yet – we will consider this topic in detail later.

How a Pivot Table Report Works

All the possibilities for further configuration are on the right pane "PivotTable Fields." At the top, by default, there is a list of all available fields, that is, columns from a flat table. The following is a section for configuration, consisting of four areas:

- Values
- Rows
- Columns
- Filters

The display of the pane can be changed by clicking the "gear" and selecting the option "Fields Section and Areas Section Side-By-Side" – so a long list of fields will be visible.

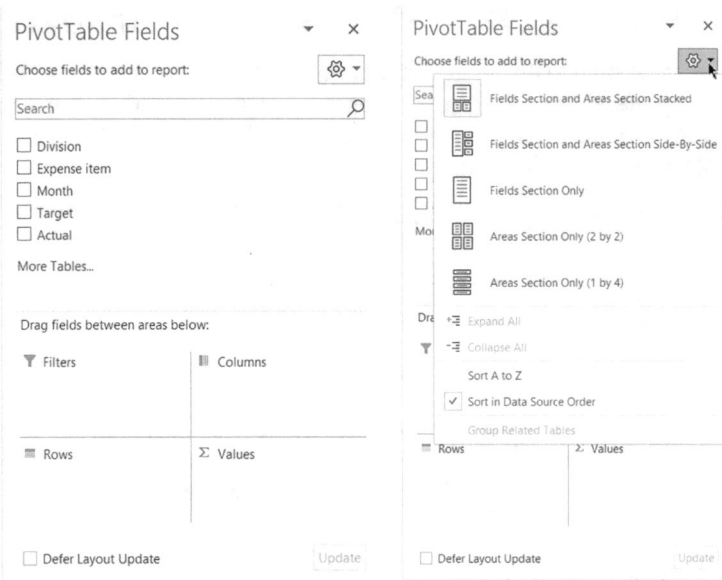

To add a field to the sample, check the box next to its name in the list of fields or drag the mouse from there to the desired area: "Filters," "Columns," "Rows," or "Values." You can also delete a field from the sample: remove the check mark next to its name or drag the mouse from a specific area to the list of fields.

The pivot table automatically aggregates data in columns, that is, combines them by some attribute. If all the data in the selected field has a numeric format, Excel will assign the summation by default. If there is at least one text or empty cell in them, then the number of cells will be counted instead of the sum.

Other calculation methods can be used in the sample cells: average, minimum, fraction, etc. There are several ways to change the calculation method.

Solution 1. Right-click on any cell of the required field in the pivot table and select another aggregation method.

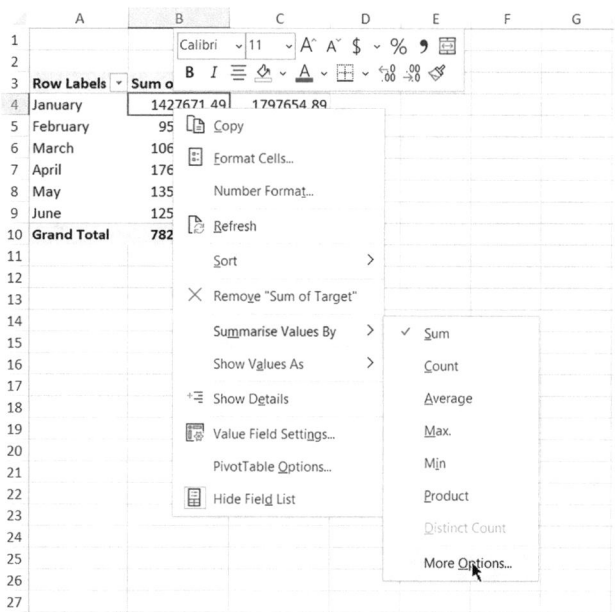

Solution 2. Select the desired type of aggregation via the menu: "PivotTable Analyse" ➤ "Active field" ➤ "Field Settings" ➤ "Summarise Values By" tab in the window that opens.

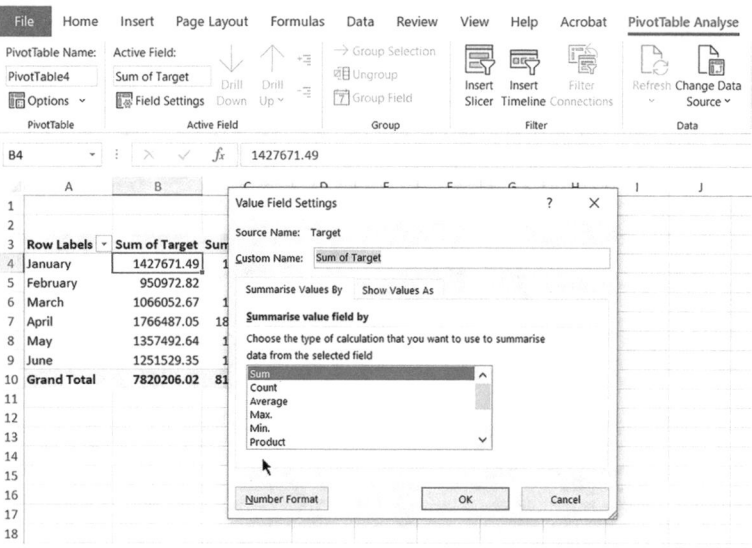

When the sample is created and configured, we make a blank for the visual element based on it.

Preparing the First Sample

Start with a sample for visualizing the dynamics of payments. On the sheet with the pivot table in the "PivotTable Fields" pane, tick the required fields: "Month," "Actual," and "Target." That's all – the first sample is ready.

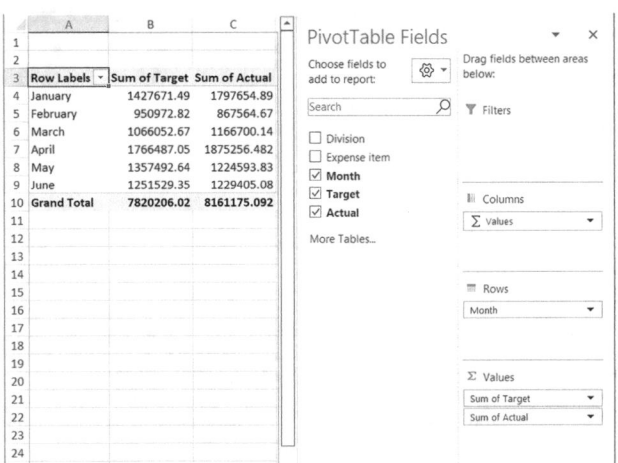

Being in any cell of the created sample, go to the "Insert" tab, click the "Insert line or area chart" button and select the "Line with Markers" representation.

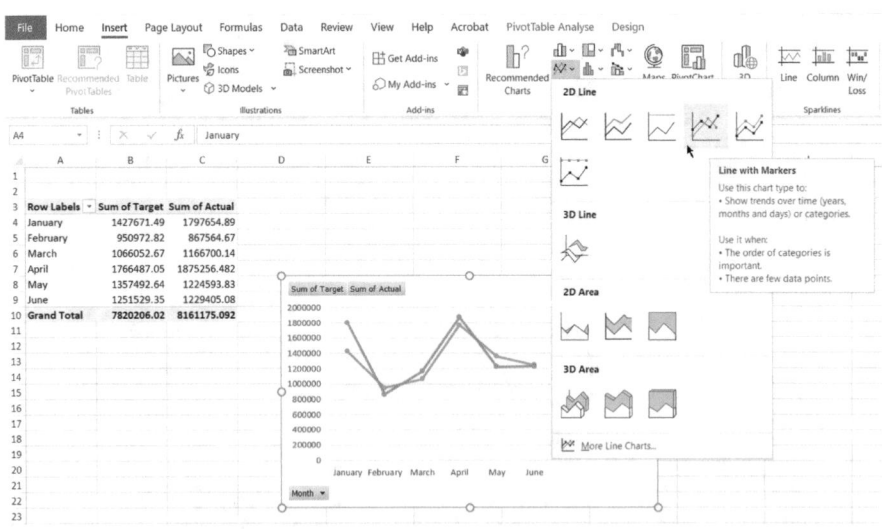

■ *Tip!* *If in the process of creating a sample something went wrong, you can always return to the original state by clicking the "Clear Pivot table" button on the "PivotTable Analyse" tab.*

In order not to get confused, give the names for the samples.

Method 1. Go to the "PivotTable Analyse" menu, click the "Options" button on the left, in the window that opens, enter the name of the table: "Payments dynamics."

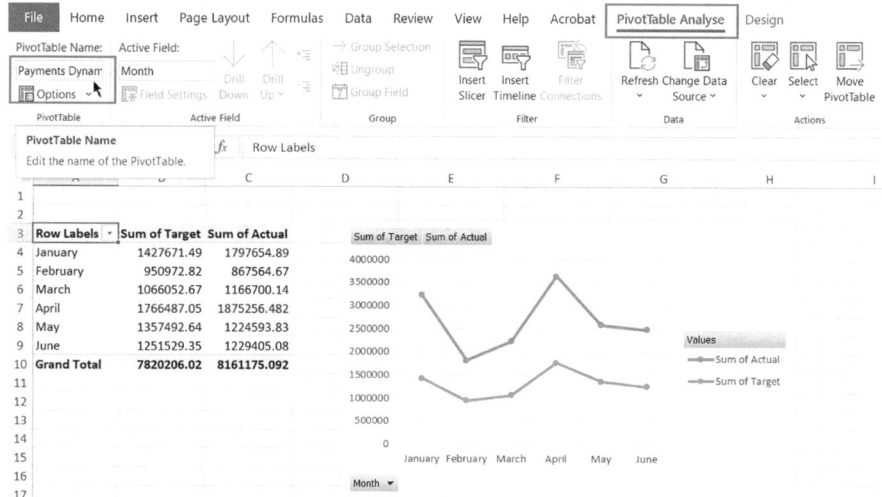

Method 2. Right-click on the first cell in the row with headers and select "PivotTable Options…" from the context menu.

In the opened dialog box in the "PivotTable Name" field, enter a name that will display the essence of the data of this sample. In further work, you will always know which sample you are working with.

This dialog box can also be called from the menu tab "PivotTable Analyse" ➤ "Options."

Replicating Data from a Pivot Table

We need a separate pivot table report for each chart. But to do this, you do not need to return to a sheet with a flat table and make a new pivot table. It is enough to copy the existing sample and paste it below on the empty rows of the same sheet. This is what I call replication.

	A	B	C
1			
2			
3	Row Labels ▾	Sum of Target	Sum of Actual
4	January	1427671.49	1797654.89
5	February	950972.82	867564.67
6	March	1066052.67	1166700.14
7	April	1766487.05	1875256.482
8	May	1357492.64	1224593.83
9	June	1251529.35	1229405.08
10	Grand Total	7820206.02	8161175.092
11			
12			
13			
14			
15			
16			
17			
18	Row Labels ▾	Sum of Target	Sum of Actual
19	January	1427671.49	1797654.89
20	February	950972.82	867564.67
21	March	1066052.67	1166700.14
22	April	1766487.05	1875256.482
23	May	1357492.64	1224593.83
24	June	1251529.35	1229405.08
25	Grand Total	7820206.02	8161175.092

■ **Tip!** *Make sure that you have selected and copied all the cells: otherwise you will not get a pivot table, but just a range of cells unsuitable for further work. This often happens when users forget to highlight totals or columns that did not fit into the visible area of the screen.*

To turn the copied sample into a new one, do the following:

- Put the cursor on any cell of the copied sample.
- On the "PivotTable Fields" pane, uncheck the "Month" field.
- In the same place, we tick the "Division" field.

As a result, we get a sample for the blank of the visual element "Expenses by divisions." Adding a blank to the sheet: on the "Insert" tab, click "Insert column or bar chart" and select the "Clustered column" representation.

In the same way, we repeat the process of replicating the sample to visualize "Expenses by items":

- We transfer a copy of the second finished sample to new rows, then put the cursor on any of its cells.
- Remove the check mark from the "Division" field on the "PivotTable Fields" pane.
- Add a check mark from the "Expense item" field.

The sample is ready, it remains to edit it. On the "Insert" tab, click "Insert column or bar chart" and select the "Clustered Bar" representation.

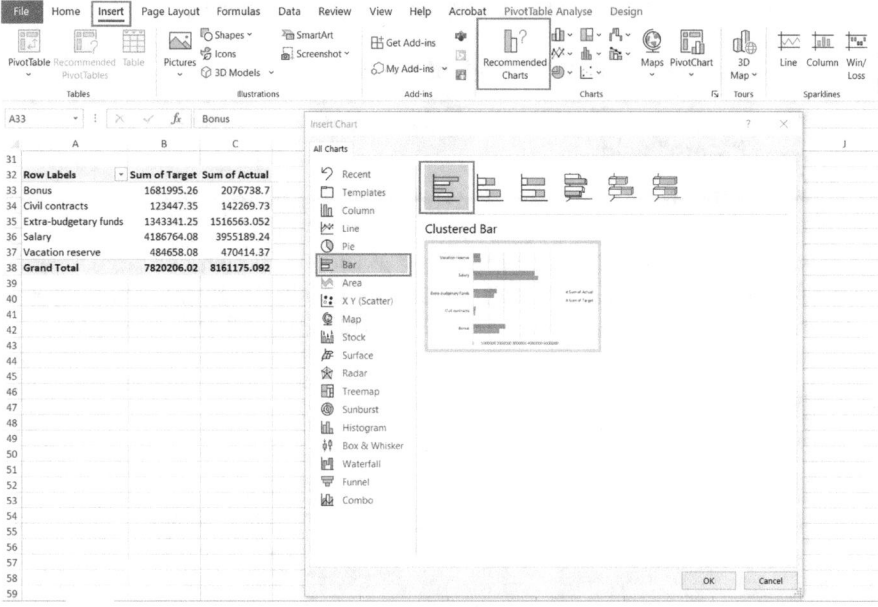

ERROR "A PIVOT TABLE REPORT CANNOT OVERLAP ANOTHER PIVOT TABLE REPORT." WHAT TO DO?

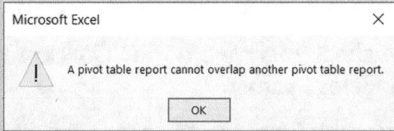

The error appears if new categories have appeared when updating the data, and they are missing rows, because the next selection is placed there.

Solution: Add the required number of empty rows (preferably with a margin) after sample with an error and update the data again.

Tip! When placing samples under each other, leave empty rows between them so that when adding data, there is enough space to automatically extend downwards. Or place them horizontally with an indentation of 1–2 columns.

Summary

All visual elements on the interactive dashboard are based on data samples from the pivot table. One element is one sample. Replication helps to save time on creating new samples: if a new one should differ from an already created one only in categories, it is not necessary to create it from scratch.

1. Select the table (make sure that all cells are selected, otherwise the selection will not work).

2. For replication, copy and paste a new sample on a sheet with a pivot table.

3. Select the fields required for the new sample in the "PivotTable Fields" pane. Give the new sample a name.

4. Standing on any cell of the sample, select and add a blank for the diagram: you will design it later.

5. Do the same with other samples and blanks of visual elements.

Leave a distance between the samples to avoid them overlapping with each other after updating the data. Determine in advance how best to place the samples: down or to the right.

1.4 Setting Up the Interactive

The key difference between a dashboard and a slide is the ability to filter data. Thanks to this, on one screen we can go from the company's totals to each division and drill down into the cost items for each month. It would take dozens of slides.

The interactivity of the dashboard can be adjusted using filters or slices. Filters are familiar to many, but they have their drawbacks that make interactions with a dashboard uncomfortable. Let's consider both tools to adjust to the preferences of any customer.

Good Old Filters

The "Filters" block is provided in the fields of the pivot table. We can move a data category there, for example, a division, and we will get a row with a filter above the table.

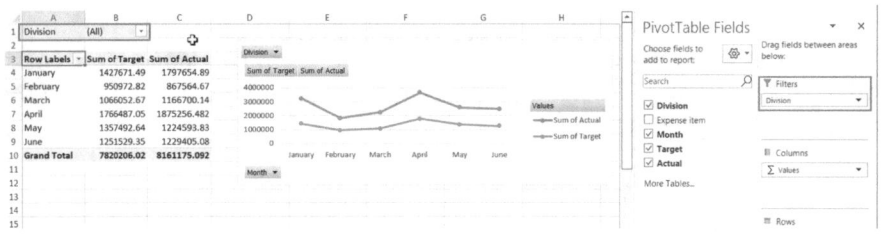

I don't like standard filters for several reasons.

- It's not immediately clear how many items are in the list: maybe four, maybe 40. You need to expand the list to select an item.

- To select several elements, you need to click a separate check mark.

- For each sample, you need to add your own filters, even if they have already been configured for another sample.

- They take up extra space on the sheet.

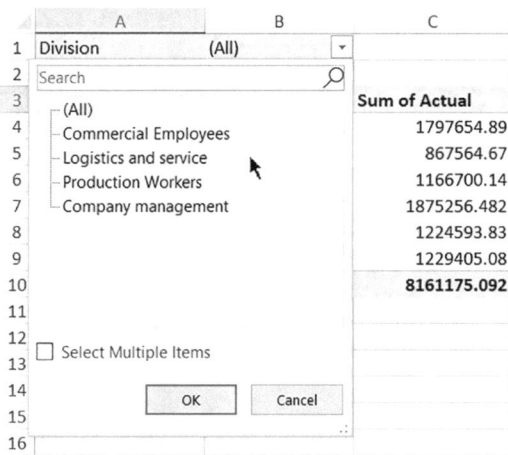

Many still add a lot of filters and tick off long lists. Although back in 2010, a modern alternative appeared – slicers. Largely thanks to them, the dashboard gets an intuitive interface.

Magic Slicers

How to create a slicer:

- Put the cursor on a cell inside the pivot table.
- On the "Insert" tab, click "Slicer".
- Check the box with the desired data category.
- We get an interactive slicer.

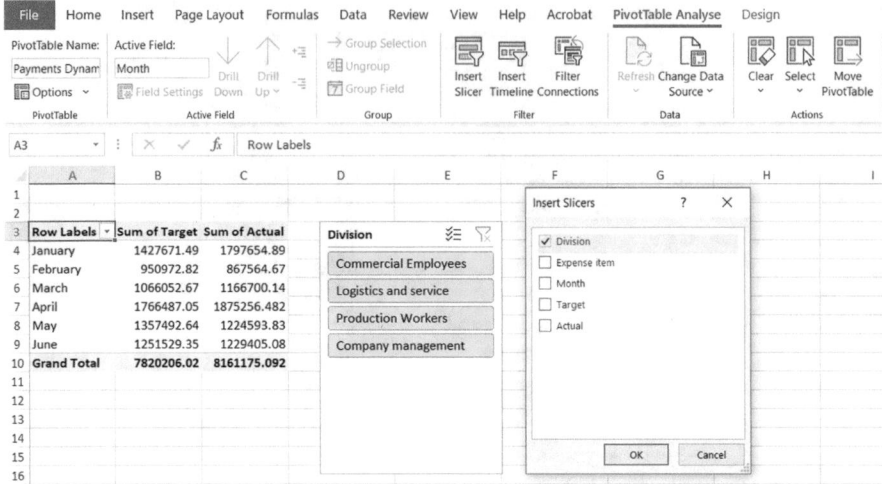

If such a window appears instead of a list of fields, it means that you missed step 1 – you need to put the cursor inside the pivot table so that the cell is highlighted. Otherwise, Excel tells you: "No connections found."

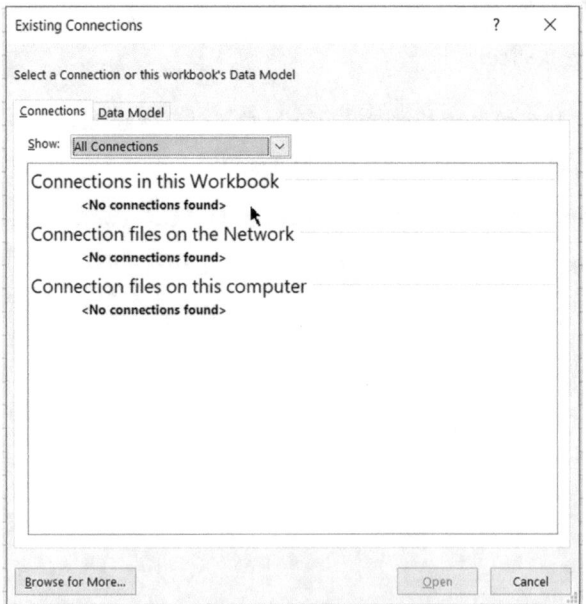

Click the slicer button – and the table together with the graph is dynamically filtered by the selected division.

If you want to select several divisions, then select them with the Ctrl (or cmd key) pressed. And if this is inconvenient for you, then you can switch to the mode of selecting several objects and select them in turn.

To reset the filter, click the familiar button at the top right.

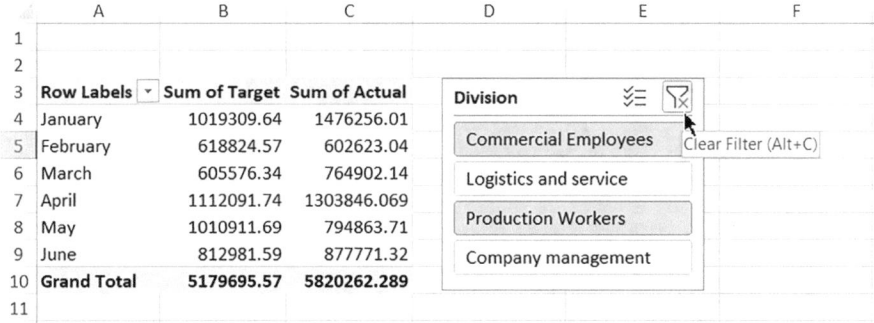

Connect Slicers to Reports

By default, the created slicer filters only the sample based on which it was created. To influence other samples, we will set up interaction.

Solution 1. Using the Button on the Menu Ribbon

On the "Slice" tab, select the "Report Connections" button. If you don't see such a tab, then your cursor is on some cell, put it on a slice.

Solution 2. Using the Context Menu

Right-click on the slicer and select "Reports Connections..." from the context menu.

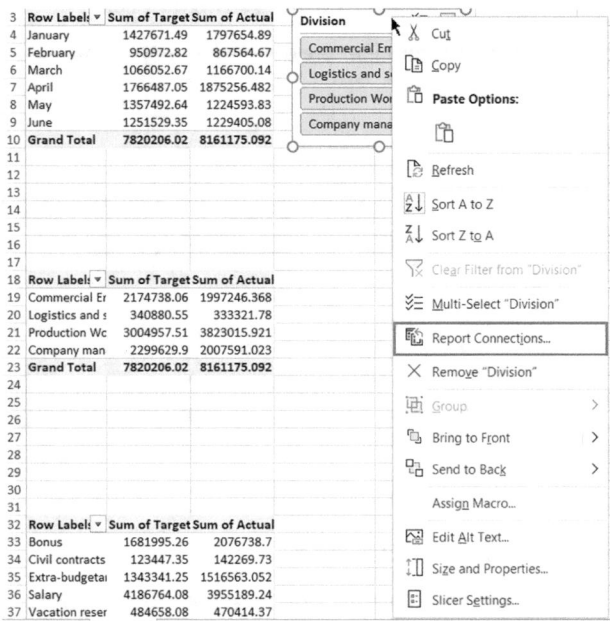

I have no preferences, I use both methods according to the situation. As a result, the "Reports Connections (Division)" window opens, where we tick all the other tables.

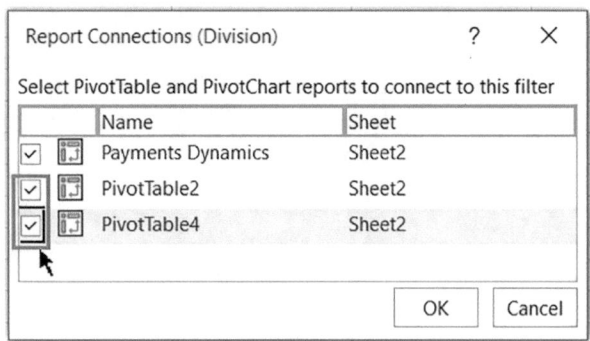

Now the "Division" slicer interacts with all the samples and blanks that are on the sheet.

You can add multiple slicers at once if you put a few ticks next to the column names in the "Insert Slicers" window.

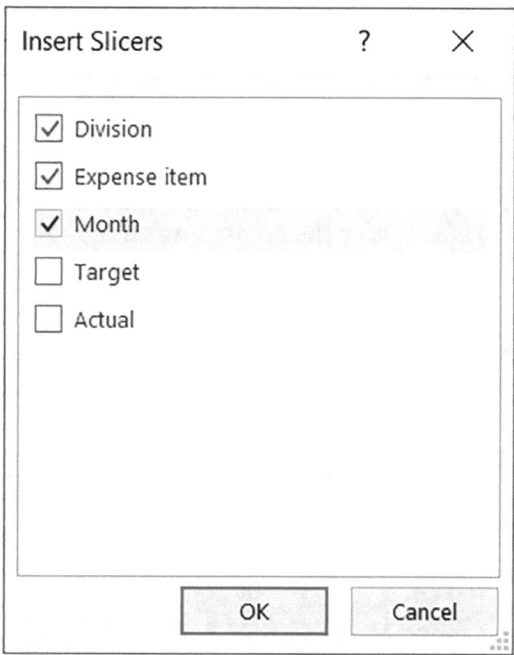

For each slicer, you need to configure interaction with existing samples. The slicer may not affect all visual elements, but only a part of them, depending on the scenario of the dashboard.

Tip! Don't create slicers for fields that contain data about actual values – use them for fields with categories.

Now the "Division" slicer interacts with all the pivot tables that are on the sheet.

The "Report Connections" button can be found not only through the "Slicer" tab but also in the context menu of the slicer that appears upon a right-click mouse operation.

You can add multiple slicers at once if in the window "Insert Slicer" put a few ticks next to the column names.

Samples and Slicers in Five Steps

Work with samples can be optimized to make the creation of visual elements easier. To do this, first we fully configure the basic sample, and then we replicate it and get an almost ready-made new one.

Step 1. Create on a new sheet the first sample with the fields of the pivot table "Month," "Target," and "Actual."

Step 2. Open the parameters of the pivot table on the tab "PivotTable Analyse" ➤ "Options" or using the context menu (right-click).

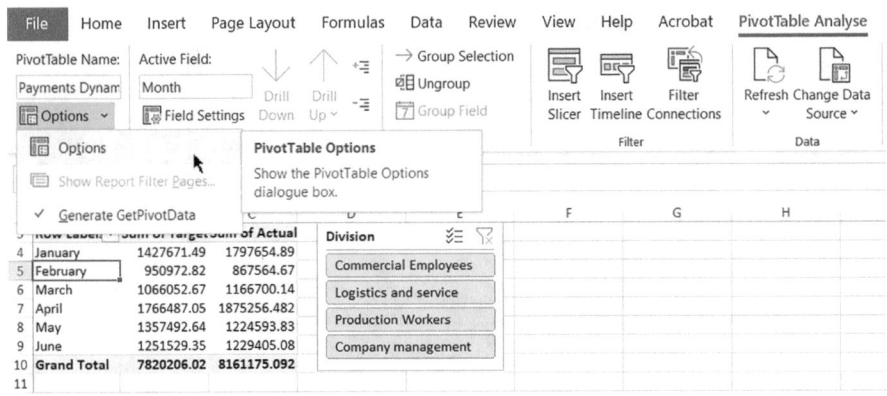

Step 3. In the settings window that opens, uncheck the "Auto-fit column widths on update" option. If this is not done, the width of the columns will have to be adjusted again each time the slicer is used.

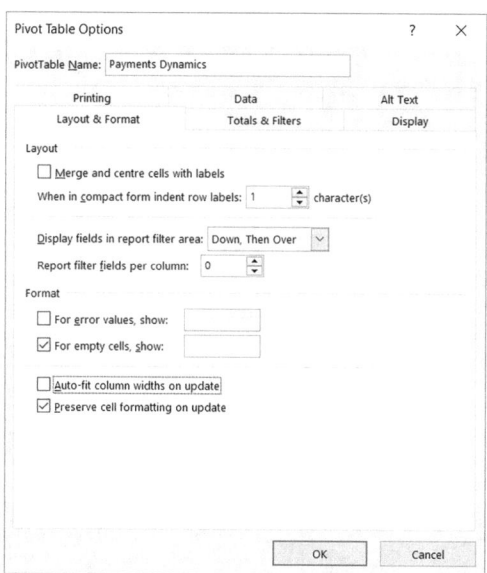

Step 4. Add the desired slicers from the base sample. On the "Insert" tab, click "Slicer" and tick the required fields in the window that appears.

To select several categories on a slicer, click on the "Multi Select" button near its title and remove the selection from unnecessary positions. Another option is to press Ctrl (or cmd) on the keyboard and select the desired positions with the mouse.

Step 5. It remains to replicate the basic customized sample and add all the necessary blanks of visual elements.

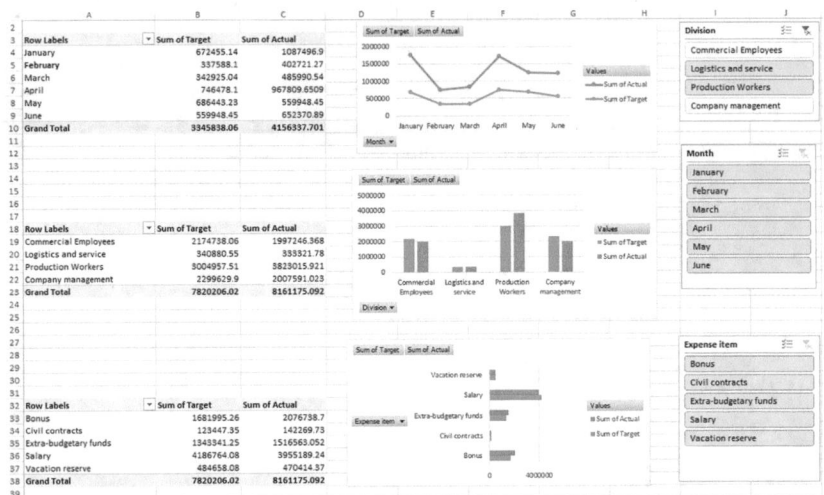

The "PivotTable Fields" pane can always be hidden, and if necessary, called up again: right-click on any cell of the pivot table and then use the "Show Field List" button.

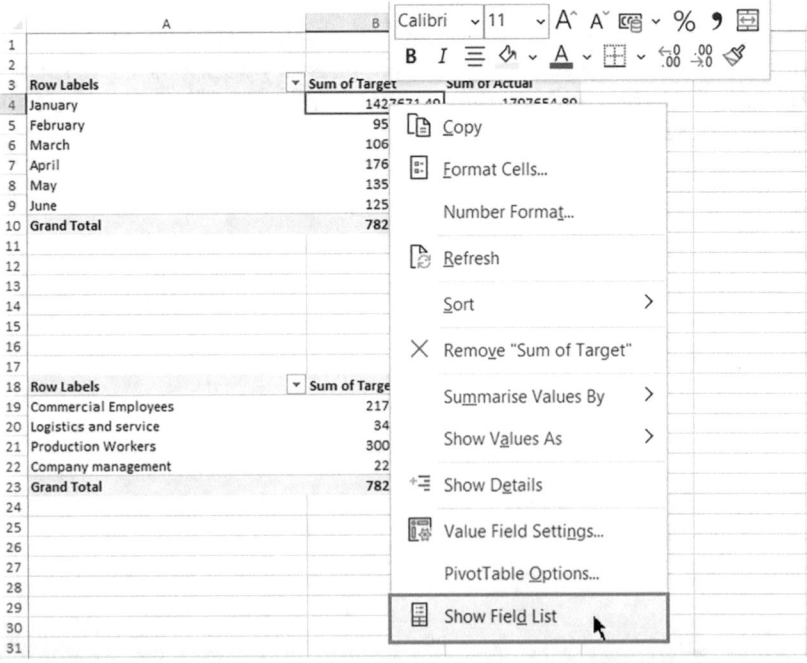

Summary

Slicers are the core of dashboard interactivity. They work like buttons, and clicking on each one updates all the data on the screen. Convenient, understandable, easy for analysis.

1. Add and configure a basic sample on the sheet with pivot table: select the required fields at the pane on the right.

2. Turn off the «Auto-fit column widths on update» option, otherwise you will have to set the column width every time after using the slicer.

3. Add a slicer from the sample ("Insert" ➤ "Slicer") and tick the necessary fields in the settings panel that opens.

4. To "deactivate" some categories on a slicer, select it, hold down Ctrl (or cmd) and click on unnecessary positions.

5. Replicate the customized sample and add blanks for visual elements.

Setting up a basic data sample before replication will greatly reduce the time to create slicers and blanks for visual elements.

Quick Tricks for Excel

"Ctrl" is used for Windows. For MacBook, you should use "cmd" instead.

"Smart" table

What happens if you convert a regular list to a table (CTRL+T):

Format as
Table ˅

- The table header is automatically frozen.
- The autofilter is turned on.
- The dimensions of the table will adjust to the data.
- All formulas will be automatically filled down for the entire column.

Quick copy down

If you need to copy the formula to the end of the column, it is better not to do it manually by holding out a black cross: just double-click on it with the left mouse button.

Filling in empty cells

To fill empty cells with values from the preceding cells

- Select the entire range.
- F5 ➤ Special… ➤ Blanks.
- Click on the "=" sign, then on the up arrow to create a link to the previous cell.
- Enter the resulting formula in all cells by pressing CTRL+Enter.

Adding a series to the chart

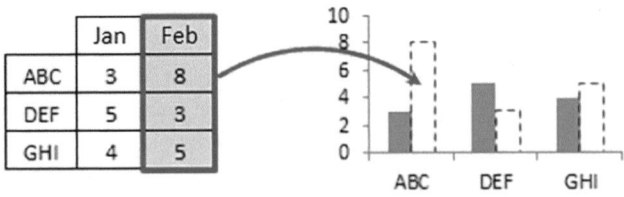

Just copy the range of cells and paste it into the chart with the right mouse button or using the keyboard shortcut CTRL+C and CTRL+V.

Dashboard Assembling

The result obtained on previous steps is only a preparatory stage. Work on the dashboard begins with the creation of its layout, which takes into account the requirements for the presentation of information as well as the location of visual elements and interactive slicers.

We have finished the data preparation, then we will work with the so-called dashboard draft. We will prepare a layout, place information blocks with visual elements on it, and assemble an interactive version.

© Alex Kolokolov 2023
A. Kolokolov, *Make Your Data Speak*,
https://doi.org/10.1007/978-1-4842-8942-6_2

2.1 Assembling Dashboard According to the Layout

You can start by creating a sketch, that is, an outline of the future dashboard. Any graphic editor, Microsoft PowerPoint program, or even a sheet of paper is enough for this – a sketch can be schematically drawn by hand. Based on this document, we will create a layout.

The structure of our dashboard can be represented as follows:

1. Dashboard title
2. Key performance indicators (KPI) cards
3. Working area
4. Slicers panel on the left

The sketch needs to be turned into a real layout, which will become our starting point. There will be more precision here – the layout should contain all visualizations, slicers, and auxiliary elements.

In the middle part (under the KPI cards), we combined two blocks into one so that we could track the dynamics of expenses over a longer period.

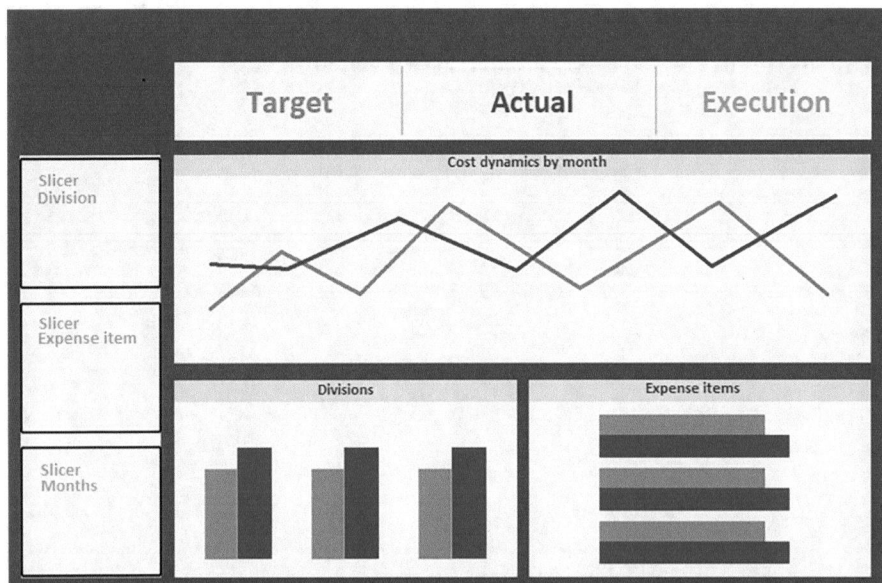

Now we have a layout and blanks for visual elements. We assemble them so that the structure corresponds to the created layout.

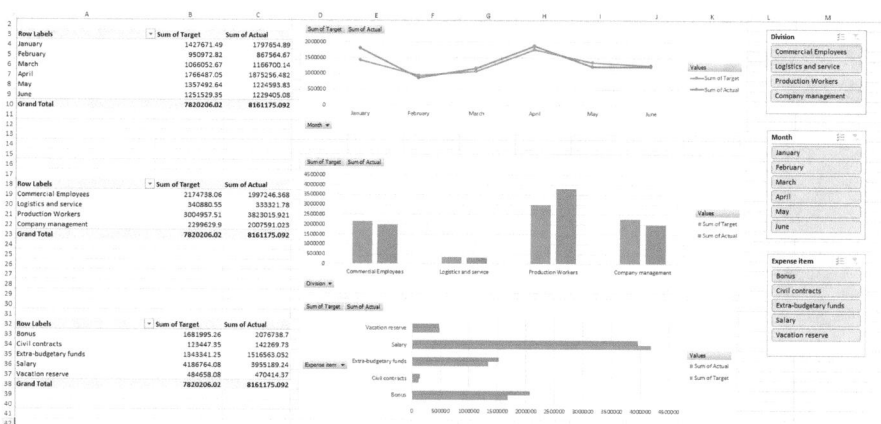

We do this on a separate sheet of our Excel file – let's call it "Dashboard."

The sheet where the blanks from the pivot table are now located will be renamed to "Draft." Here we will store and prepare all the necessary data for the final dashboard.

First, we will transfer all the slicers to the "Dashboard" sheet. To do this, select the first slice on the "Draft" sheet and, pressing Ctrl (or cmd) on the keyboard, select the rest with the mouse. We cut out the entire group of slicers.

■ **Tip!**   *Don't leave copies of visual elements on the "Draft" sheet, so as not to get confused in them in the future. Transfer everything you need for work to the "Dashboard" sheet, and leave the "Draft" for data preparation and calculations.*

Go to the "Dashboard" sheet and insert the cut. We place the slicers on the left according to the layout, leaving five lines on top for the title of the dashboard and KPI cards.

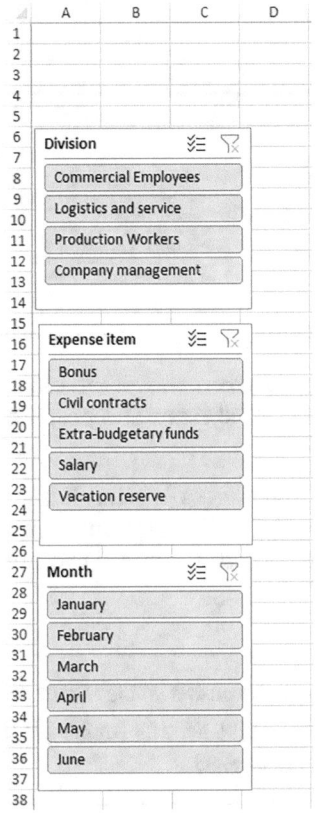

Then we transfer the prepared diagrams from the "Draft" sheet to the "Dashboard" one: the same way as we did it with slicers. We place the diagrams on the sheet in accordance with the layout:

- On top of the entire width of the screen is a graph of dynamics.

- At the bottom left is a clustered column chart for divisions.

- At the bottom right is a clustered bar chart for expenditure items.

We already have something similar to a dashboard. In this case, only the samples from the pivot table remain on the "Draft" sheet.

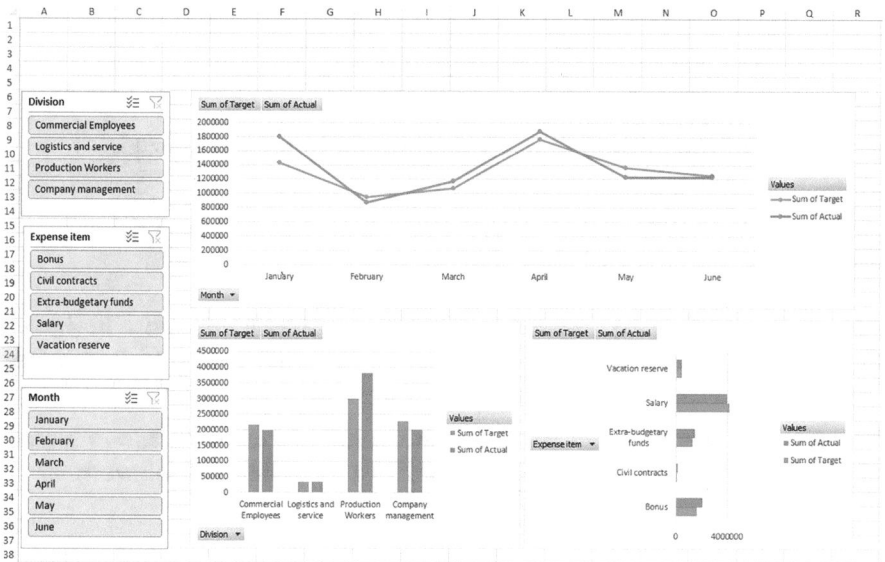

Summary

Start working on the dashboard by creating a sketch and layout. Use any visual editor for this or make a sketch by hand.

1. Determine which visualizations are needed on the dashboard.

2. Arrange them on the layout according to the principle "from general to specific."

3. Make sure that the most important data is not "lost" against the background of the rest.

Assembly according to the layout is faster and easier for most beginners. Firstly, because we made the blanks for the visual elements in advance. Secondly, because here we get a visual result of the work done.

Visualizations are not final yet. But the interaction with the slicers already filters the charts, and we see how the interactive dashboard will work.

2.2 Creating KPI Cards

At the top of the dashboard, we left space for the totals. There is no need to build diagrams here, it is enough to display the number and its title large enough. I call such an object a KPI card. We need to create three such cards:

- Target
- Actual
- Execution

To do this, you will also need data samples from the pivot table. They can be replicated in the same way as we have already done for blanks of visual elements. Already created samples are not used here, since additional calculations or transformations may be required for KPI cards.

Sampling Data for Cards

To replicate, copy any sample on the "Draft" sheet, paste it a little lower and on the "PivotTable Fields" pane, uncheck the box next to the name of the field with the "Expense item" category. You can also remove this field from the "Rows" area.

So we get a sample on a pivot table with data on key indicators with a target and an actual.

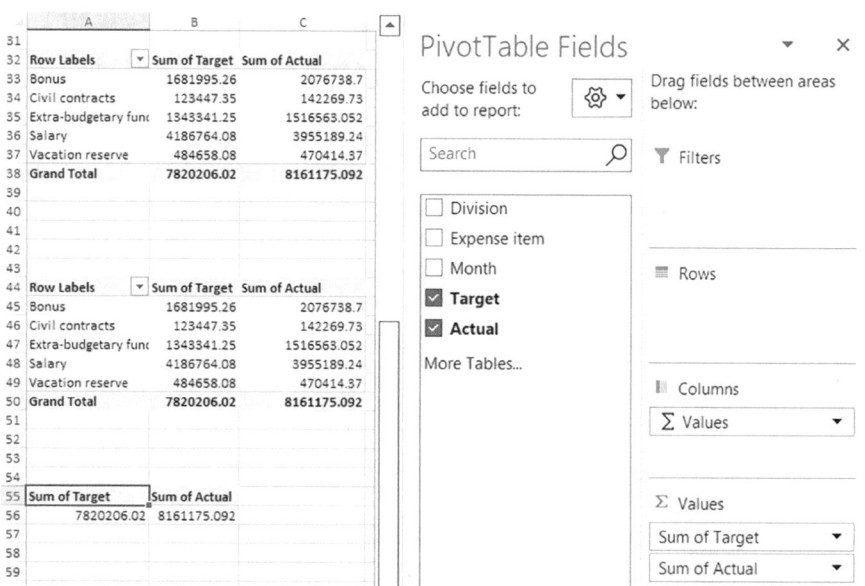

The "Target" Card

On the Dashboard sheet, we start creating the first KPI card. Putting the cursor in the cell, we add a link to the general indicator of the target on the "Draft" sheet. To do this, enter = Draft!A56, where A56 is the address of the cell on the sheet with samples. We get the required value already on the Dashboard sheet.

C2		:	✕	✓	*fx*	=Draft!A56		
	A	B	C	D	E	F	G	
1								
2			7820206					
3								
4								

IF THE LINK LOOKS WRONG, DISABLE THE GETPIVOTDATA FUNCTION

Initially, the cell link may look like this:

=GETPIVOTDATA("Sum of Target";Draft!$A$56). In this case, you need to disable the GETPIVOTDATA function.

This function is useful: it extracts the necessary data from the pivot table to use it in other tables or calculations. But when creating a dashboard, it becomes a headache.

Being in any cell of the sample, go to the tab "Pivot table Analyse" ➤ "Pivot table" ➤ "Options" ➤ uncheck "Generate GetPivotData."

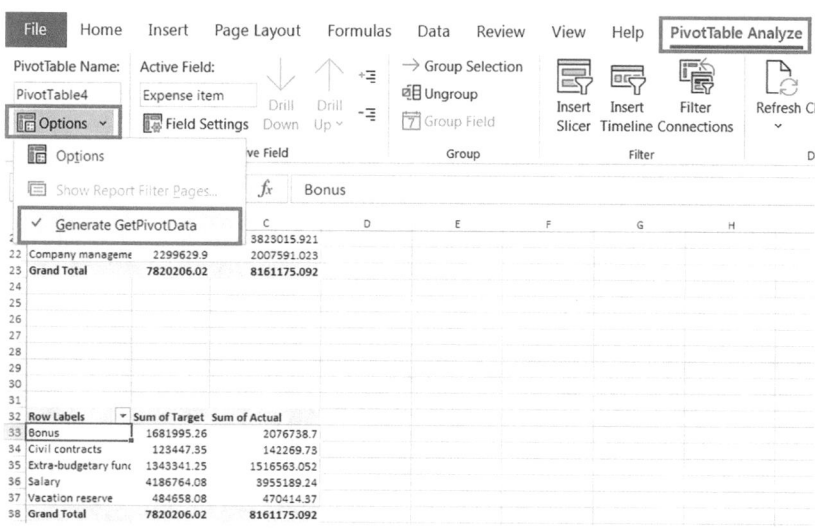

So you will get cells' links of the pivot table not using GETPIVOTDATA, but in the usual way, in the form of links to cells A1, B2, or C4.

Note that if the links to the pivot table were created before disabling the function, they remain in their original form after disabling.

■ **Tip!**　*If the number in the card consists of a large number of characters, change its bit depth. For example, convert into thousands or millions. This way the indicator will become more compact and easy to understand.*

We give a name to the indicator and set its bit depth "Target, $K". The initial data in the cell is divided by 1000. It remains only to set the type of number. We do this on the "Home" tab in the "Number" section:

- Clicking on the "Comma style" icon (looks like a comma) will format number with a thousand.

- Clicking on the icon next to "Decrease decimal" (000 with the right arrow) will show fewer decimal places.

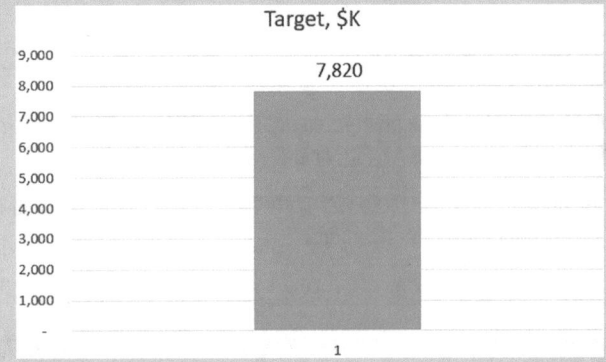

DON'T BUILD A CHART FOR THE SAKE OF A SINGLE NUMBER!

Such visualization does not add informativeness, but distracts attention from the number and forms information noise in the form of a scale, lines, and columns. There is nothing to compare one column with, so there is no point in it – present the KPIs in the cards.

To focus on the value of the indicator, we will increase its font to 20 pt.

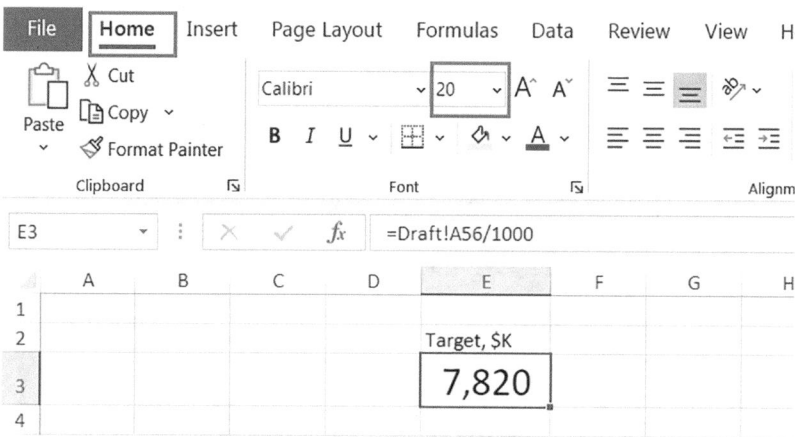

The fill will help to focus attention on the card: add it to the cell with the number and the name of the indicator. Do not choose dark or bright colors, otherwise the values may be lost on their background.

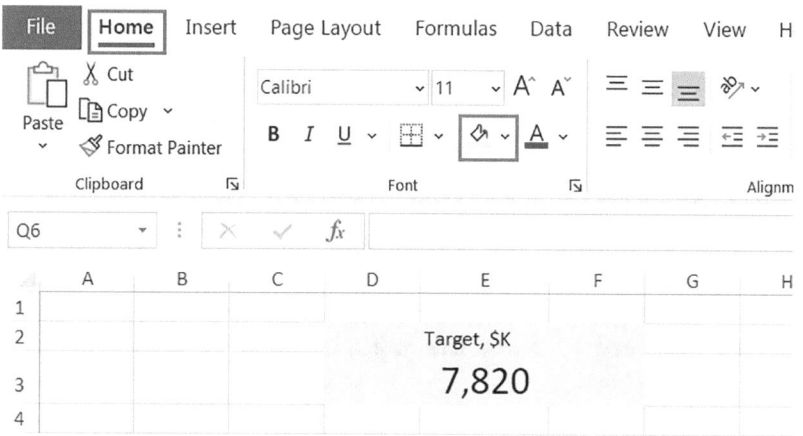

To select the card size, stretch the column with the value. In this case, the alignment of the number may be disrupted. To fix this, we will set the cell format on the "Home" tab: in the "Number" section, open the dialog window by clicking on the rectangle with an arrow. In the "Cell Format" window that opens, you must

- Select the "Number" category in the "Number" section.

- Make sure that the number of decimal places is zero.

- Make sure that there is a check mark in the "Use 1000 Separator (,)" field.
- Click OK.

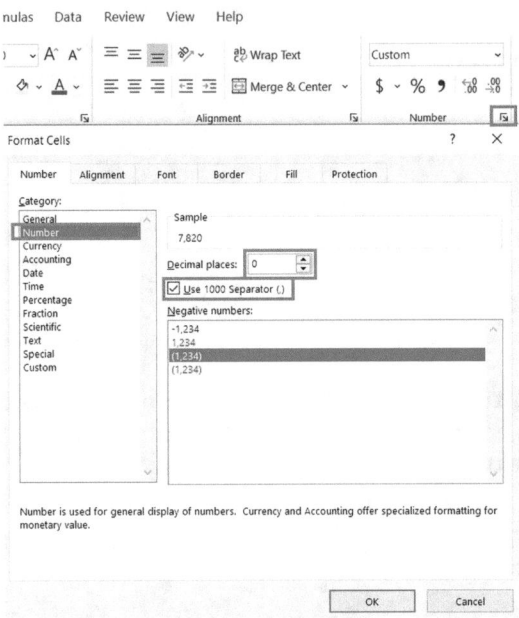

That's all – the card with the target indicator is ready.

"Actual" and "Execution" cards: creation and configuration.

Then we simply replicate the created card: copy, step back two columns, leaving a gap, and paste. We change the link to the cell in the copied card (instead of the target indicator, we now need the actual one). As in the first card, here you need to divide the indicator by 1000: make sure that this remains in the formula.

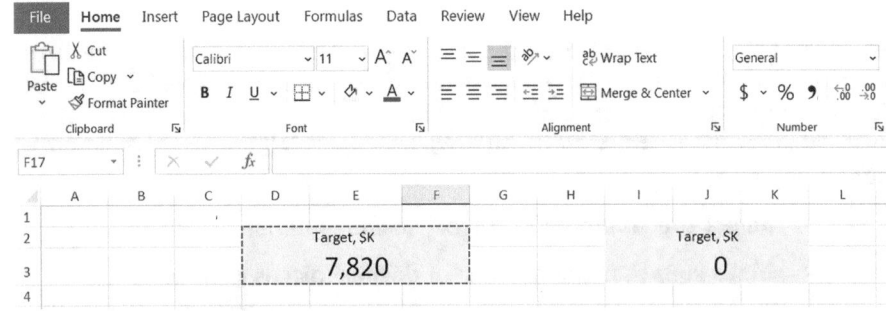

We change the title of the new card to "Actual, $K". It's not a big deal if the data doesn't fit and is displayed in the cell as "#######". To see the actual value, just stretch the column width. It remains only to bring the received data to the desired bit depth.

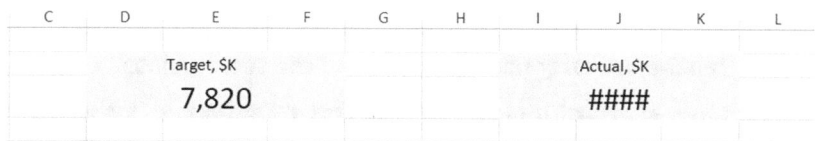

The second card "Actual, $K" has been created. Since we copied an already configured card, we didn't have to repeat all the configuration steps.

To create a third "Execution" card, copy one of the created ones and paste it also at a distance of two columns from the neighboring one. To calculate the value in this card, it is enough to insert a formula dividing the value from the "Actual, $K" card by the value from the "Target, $K" card.

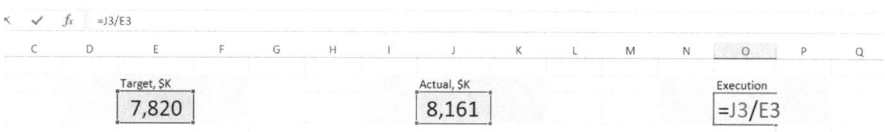

The resulting value is displayed as a percentage; to do this on the "Home" tab in the "Number" section, click the "Percent style" button. We give the created cards visually the same size – and they are ready.

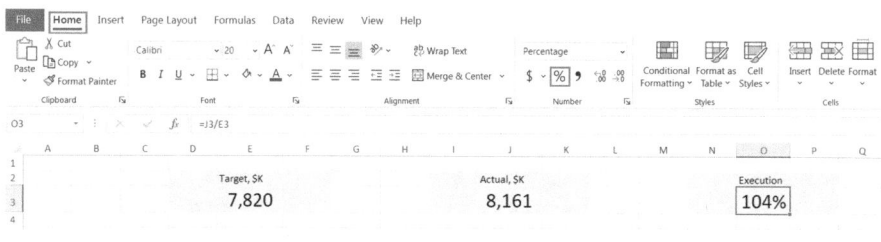

■ **Tip!** *Don't overdo it with the number of KPI cards. If the customer asks to display 7–10 indicators on them, explain that this will make the numbers difficult to perceive and will lead to an overload of the dashboard. I recommend adding no more than six cards.*

Summary

KPI cards are the first thing the user sees on the dashboard. There is no need for diagrams and bright color, all attention should be focused on an important numerical indicator. We created these elements in a few steps.

- On a sheet with pivot tables, we formed a sample with general indicators of the target and the actual values.

- On a sheet with blanks of diagrams ("Dashboard"), the first card was designed: it referred to a cell with a target indicator on another sheet. We adjusted the bit depth for the card and gave it a clear title.

- We designed the appearance of the first card: we increased the font of the numerical indicator and added a light blue fill to focus attention on the KPI.

- By replicating an already-created card, two new cards, "Actual" and "Execution," were created. In this case, you will not have to repeat all the configuration steps, but you need to change the formula for the indicator of the "Execution" card.

2.3 Align the Dashboard and Add a Header

Due to the change in the size of the cards, the diagrams on the layout stretched, and some of the data on them went off the screen. But it is important for us that the user sees all the information at once.

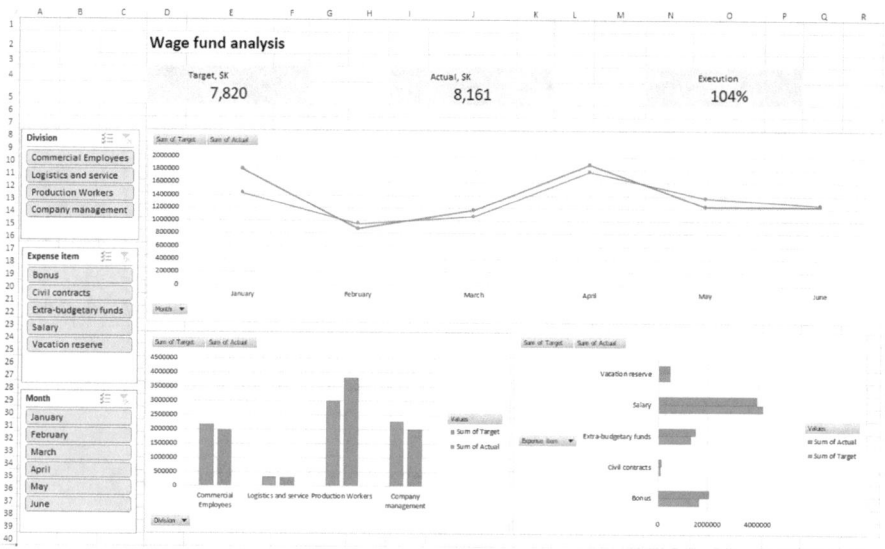

To fit everything on one screen, you need to change the sizes of graphs and charts. To do this, select a block with a diagram and adjust the size by moving the lines of its borders.

We do the same with the rest of the visual elements. If you need to move the block with the diagram exactly horizontally or vertically, select it and, holding down the Shift key, move it with the mouse.

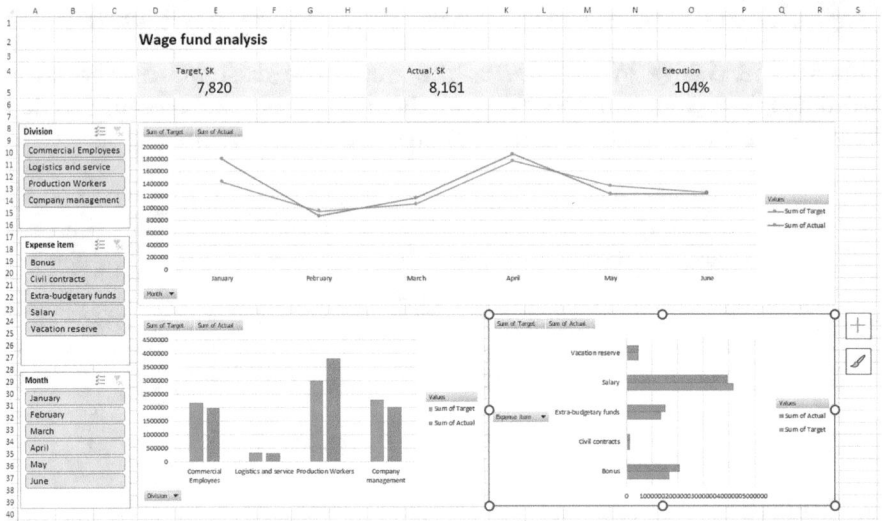

Remove the extra row between the cards and the visual elements. Now let's add the dashboard title to the sheet.

- We need three empty rows above the KPI cards: add if there are not enough of them.

- In the second empty row, in the cell above the first card (in our case it's D2), write the title of the dashboard "Wage fund analysis."

- Set the font size of the title to 20 points: the same as the numerical indicators in the cards.

To complete the work on the dashboard at this stage, it remains only to remove the grid on the dashboard sheet. To do this, just one mouse click is needed: on the "View" menu tab, uncheck the "Gridlines" parameter.

This is what our dashboard looks like after all the assembly operations.

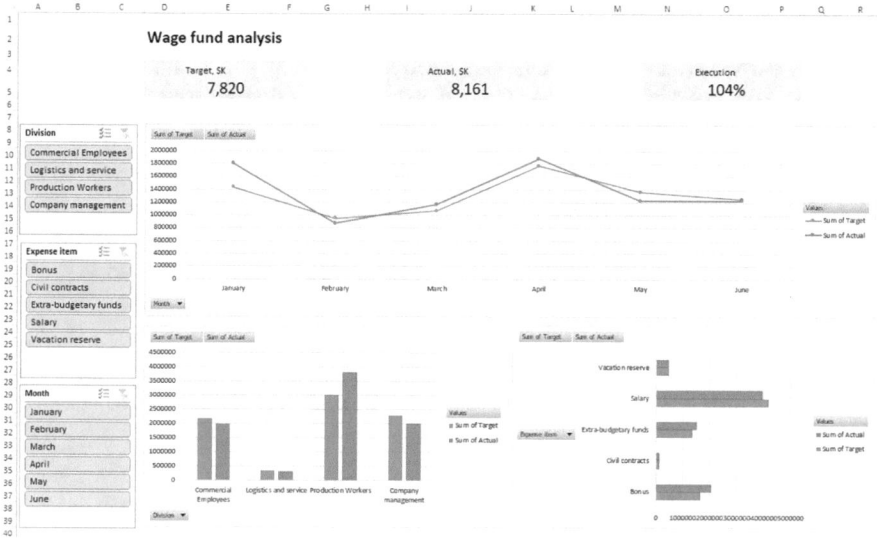

Summary

Practice shows that any carelessness reduces not only the value of your work but even the credibility of the data. A well-designed dashboard costs more. Therefore, it is important to pay attention to the proper arrangement of elements and their alignment.

Here's what we did for this:

- We changed the size of the charts to fit everything on one screen
- Aligned the location of visual elements
- Added the dashboard title to the empty lines above the cards
- Removed the gridlines from the Excel sheet

Anatomy of Diagrams

We built a dashboard pretty quickly. And it looks like an interactive analytical application, not just a set of charts and filters, as it usually turns out in Excel. It seems to many that it is already good, but we are only at the beginning of the journey.

This part of the book is about fine-tuning the elements of the diagram, from which the first impression is "beautiful and clear." You don't have to be a designer to make your work look professional and expensive. You just need to follow a number of rules, which I will tell you about.

These rules are relevant for any tool, be it Excel, PowerPoint, Power BI, Tableau, or some other software. The names of the chart elements in them may differ, but the principles of configuration are the same. What you will learn to do in an ordinary MS Excel you will be able to apply competently in advanced programs.

© Alex Kolokolov 2023
A. Kolokolov, *Make Your Data Speak*,
https://doi.org/10.1007/978-1-4842-8942-6_3

3.1 Analyzing Ready-made Design Styles

Excel offers a whole set of templates to customize the appearance of charts. Let's figure out what they are convenient for, and what they are not suitable for.

To view the suggested options, select any diagram and open the "Design" tab that appears in the menu. The list of standard templates can be seen in the "Chart Styles" section and in the drop-down menu under the "Quick Layout" button.

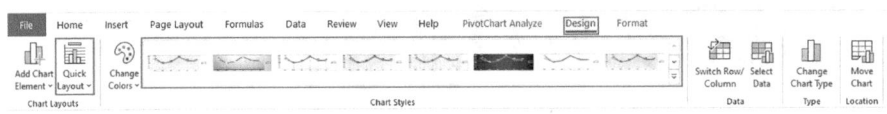

How to Choose a Quick Layout

Excel offers 12 template options for our "Payments dynamics" chart. They differ in the presence or absence of some elements, as well as their different location.

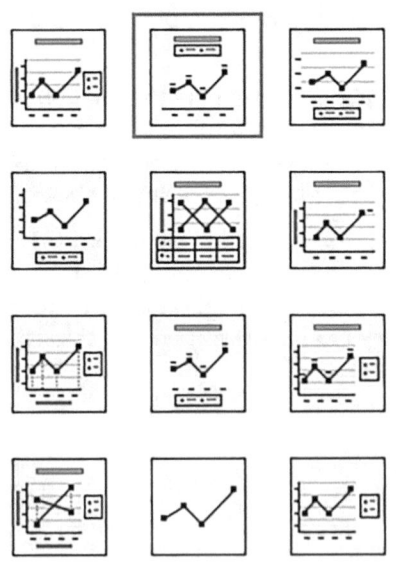

Of these options, "Layout 2" is more or less suitable for us, because it has the necessary set of data and elements:

- Chart title.
- Correctly positioned legend (above the chart).
- Data labels.
- Horizontal axis only (vertical is not needed because there are data labels).
- There are no gridlines so as not to create barriers to the perception of information.

The rest of the quick layouts are not suitable for our tasks: they either do not have the necessary elements, or there are a lot of unnecessary ones. Therefore, when there is no time, such a layout can be taken for further improvement.

We will not use this option, but we will fine-tune the diagram ourselves so that you understand how this process works. But first let's see what templates Excel offers in the "Chart Styles" section.

What Is Wrong with Template Styles

In each of the suggested styles, certain settings have already been done: the background of the diagram, the location of the legend, font sizes, visual design of lines or columns.

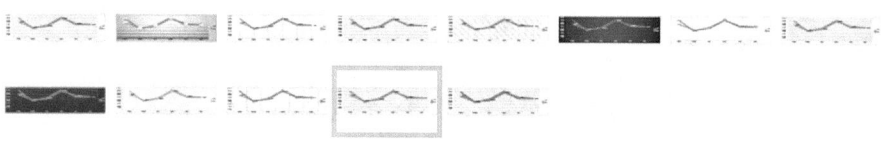

Let's look at several suggested options and figure out why they don't suit us and what mistakes in the design should be avoided.

A template with an emphasis on data. As a first example, let's consider a variant of the style with bright large balls.

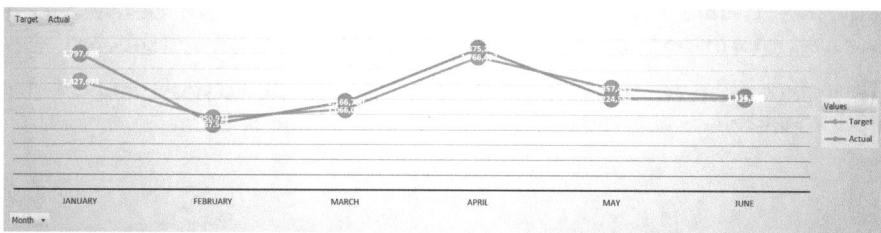

The balls draw attention to the data, and this makes the graph original. But the template still has drawbacks.

A background gray gradient brightens the top of the graph and darkens the bottom. For business graphics, such decorations have no practical meaning. In addition, white data labels are completely lost against this background.

There are also questions about the grid. The lines from the almost imperceptible ones at the top become wider closer to the base – this adds "gravity" to the graph. And in general, the grid does not perform its functions here. If there are no labels, it helps to correlate data points with values on the Y axis. But there are labels here, but there is no Y axis.

Template on a dark background. To get a complete picture of the available styles, we will not limit ourselves to templates for the "Payments dynamics" chart – we will consider an option for a chart with expenses by division.

The style template on a dark background attracts attention with its "neon," as if glowing columns. But for practical use in our case, it is not suitable.

Firstly, there are no data labels here – we don't see specific numeric values. Secondly, the gridlines "blur" closer to the base – this makes it difficult to correlate small columns with values on the Y axis. And thirdly, it is unclear why the columns need to "glow." After all, our goal is to clearly convey business information, and not to simulate a dance floor on a dashboard.

Template with vertical labels. Among the suggested styles, there are also options without a colored background. Consider one of them.

This style looks good, but it also has too many disadvantages to use on a dashboard.

Vertical data labels are terrible. To read them, the user will have to tilt his head, and this definitely does not help the comfortable perception of the report.

The columns are very thin and the distance between them is very wide. This does not give us any advantages, it only forms a lot of empty space on the diagram.

A separate story is the captions of categories in capital letters. And it's not just that they take up more space and create unnecessary emphasis. It's just that whole phrases in upper case create a feeling of "screaming," and therefore are considered a sign of bad taste.

Capital letters in abbreviations or short names are normal, the main thing is not to overdo it with accents. Just such an example will be in the following chart style template.

A template with a gradient on the columns. To complete the picture, let's also consider a variant of the standard style for a bar chart.

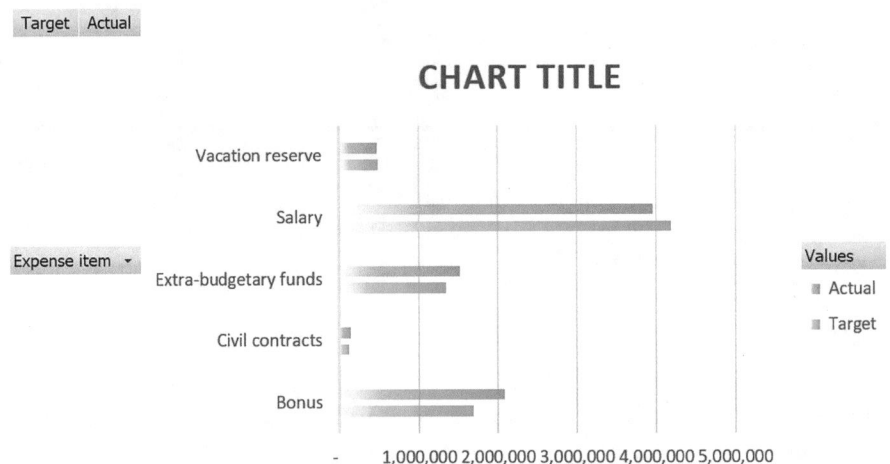

The first thing that catches the eye is the title of the diagram. I will not criticize the capital letters – they could be appropriate. But. Here it is the most noticeable element, and it distracts attention from the data.

There are three tools for placing accents:

- Font size
- Bold
- Capital letters

It is quite enough to choose one of them, rather than use them all at once. Otherwise, you will get an explicit overkill, as in this example.

The design of the columns raises no less questions. First of all, they are very thin: that's also why all the attention goes to the title. Secondly, the gradient on them looks superfluous decoration – there is no practical sense in it.

Also, this style template does not provide data labels, and the user does not see the exact numeric values. The grid helps a little to correlate the length of the columns with the labels on the X-axis, but its vertical lines seem to divide these columns into parts and violate their "integrity."

What Should Be the Business Graphics: Recommended Samples

The diagrams look different in the proposed templates and styles, but none of the options is suitable for our purposes. In the following parts, we will customize the desired style ourselves and create a minimalistic diagram template for solving business problems.

The graph will have the necessary minimum: title, legend, and data labels. We will disable the auxiliary elements that Excel offers by default in order to focus on the main data.

We will also put the bar charts in order. We will leave the necessary minimum, but we will remove the gap between the columns on the chart so that the target and actual indicators of each division look like a single block.

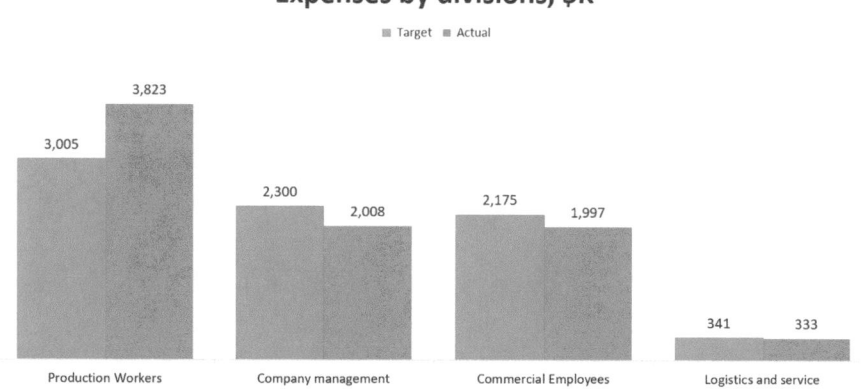

Let's also deal with the design features of the bar chart "Expenses by items," we will build the correct sorting of columns on it: Excel has its own surprises here.

Summary

By default, Excel builds "rough" versions of charts, and offers a set of layouts and styles for their design. We checked some of them to see if these options are suitable for us.

- We examined the proposed express layouts on the "Design" tab: the required data set contained only one of them.

- We made sure that this single express layout also needs additional settings.

- We tested several standard chart styles, noted the pros and cons of each.

According to the results of such a test, it became clear that none of the proposed standard options would suit us without improvement. Therefore, we will continue to delve deeper into the anatomy of diagrams and add fine-tuning ourselves.

3.2 Setting Up Data Labels

According to our graph, it is already approximately clear how the planned and actual payments have changed over time. It can be seen that there was a slight increase in expenses in April. Or that the actual line slightly exceeds the target.

But it is impossible to discern the exact value here. For example, how much did the company spend in May? This requires data labels. At the same time, it is important to make them visual: it happens that the labels seem to be there, but it is very difficult to see these small numbers against the background of other elements.

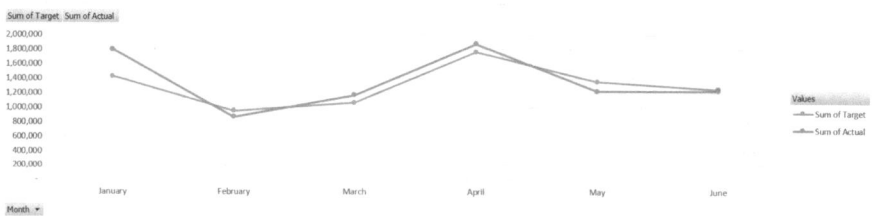

How to Add Data Labels

First of all, we will add the data values to the graph. There are three ways to do this.

Solution 1. Select the block with the graph and click "Add chart element" on the "Design" menu tab. From the drop-down list, select the sub-item "Data labels" and specify exactly where you want to place them.

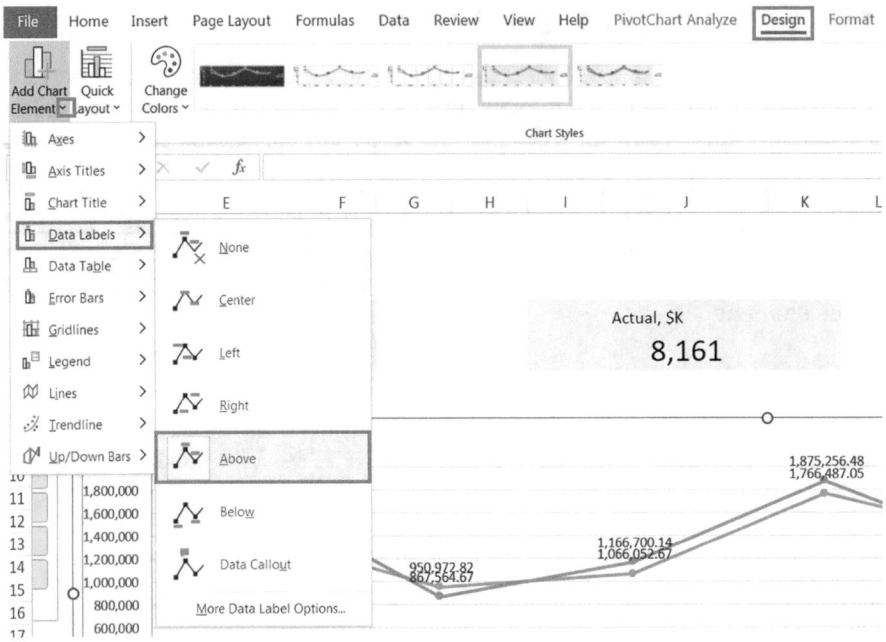

Solution 2. Select the block with the graph and click on the icon on the right in the form of a green cross. In the "Chart Elements" menu that appears, check the "Data Labels" box and select their location at the top.

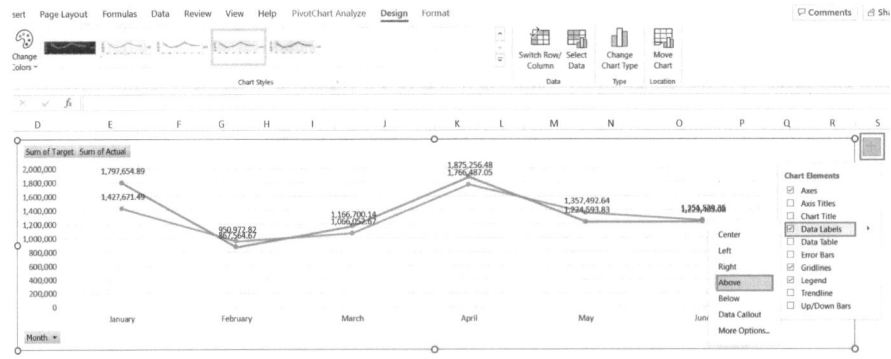

Solution 3. Select the graph line by clicking on it and open the context menu with the right mouse button. Select "Add Data Labels" ➤ "Add Data Labels."

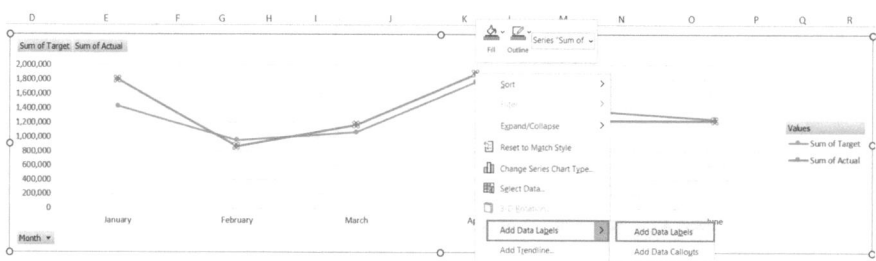

All these methods lead to the same result, but I use the second one, where data labels are added by clicking on the "+" icon to the right of the chart.

Labels Format: Getting Rid of Long Numbers

Data labels have appeared. But there are nine figures in each number: it is difficult to perceive this, so the number of figures in the numbers should be reduced.

First, let's change the bit depth of the Y axis from units to thousands.

1. Select the values on the Y axis with a mouse click.

2. Right-click the context menu and select "Format Axis …"

3. In the panel that opens on the right near the "Display units" field, select "Thousands" from the drop-down list.

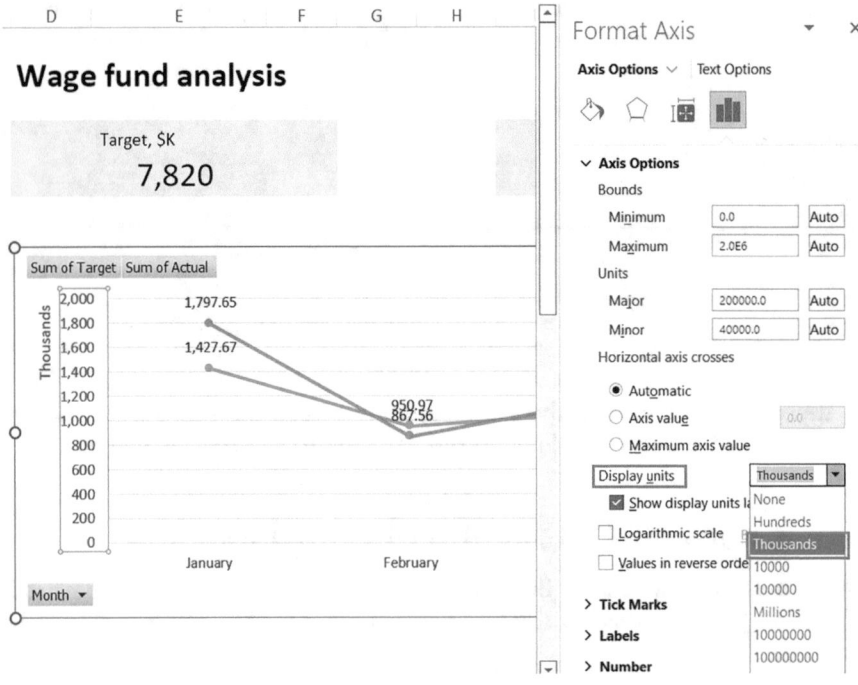

We set the bit depth of the axis, but the same nine figures remained in the labels. To fix this, we will also change the format of the data labels.

1. Select a series of data on the graph with one click on any number. To select one number, you need to click on it twice.

2. Open the context menu with the right mouse button and select "Format Data Labels…"

3. In the panel that opens on the right, in the "Number" section, set the "Number" category.

4. In the "Decimal places" field, specify "0" and tick the checkbox next to "Use 1000 Separator (,)."

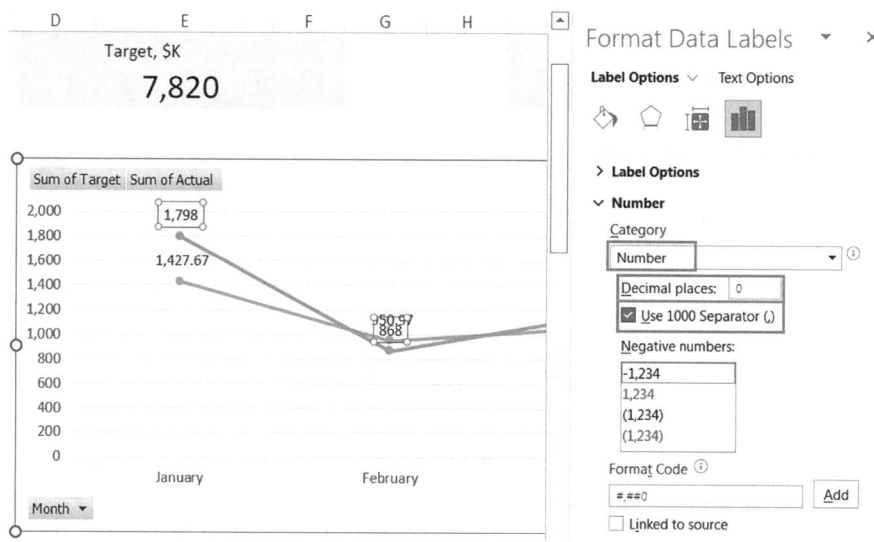

The selected data series has acquired the desired form – the numbers are well readable, they are easy to perceive.

HOW TO FORMAT LABELS THROUGH DATA IN A TABLE

Go to the "Draft" sheet and select the desired data range. On the "Home" menu tab in the "Number" section, format the type of numbers using the "Format with thousand separators" buttons (the "'" icon) and "Decrease decimal" ("000" with an arrow).

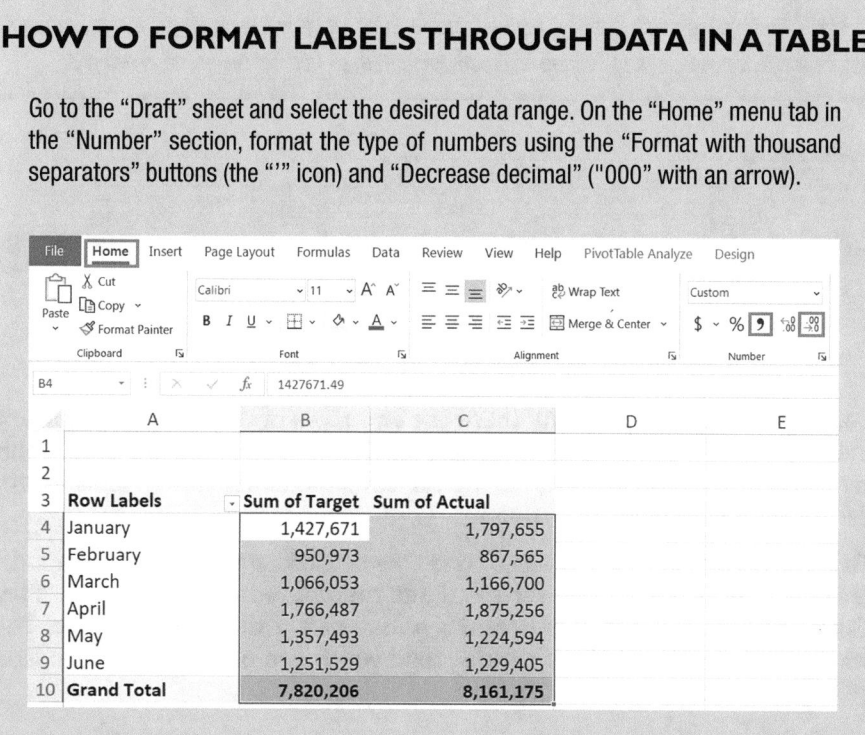

After these actions, all data labels on the chart are displayed in the desired format.

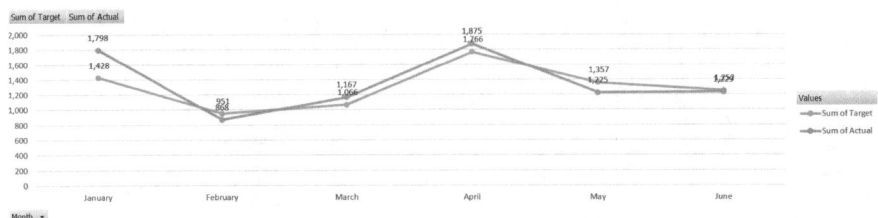

Size and Color for Data Labels

By default, Excel has set size 9 for the font of data labels. To make it easier to read and perceive them, we will increase it to 12.

Select the block with the diagram and specify the desired size on the "Home" tab in the "Font" section. Automatically, the font will increase on several elements at once: in data labels, in axis labels, and in the legend. All these changes completely suit us.

■ **Tip!**   *Some clients ask to add a background for data labels because they are "so used to it." But for modern styles, this technique is already outdated, so I do not recommend using it.*

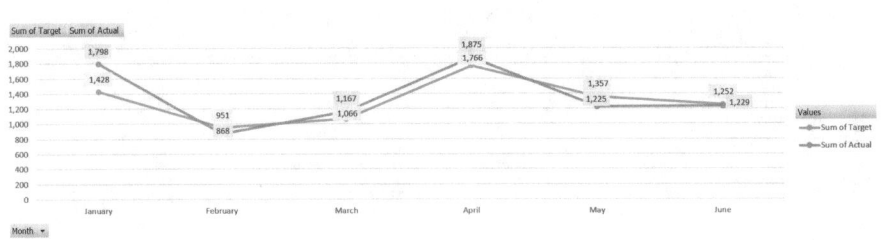

But there is another problem: there are two rows of data on our graph, and their labels overlap each other. Excel does not have a button that would eliminate such a disadvantage, but we can compensate for this by adjusting the data labels to the color of the row.

To do this, I select all the labels of one row by clicking the mouse and on the "Home" tab in the "Font" section, I set the desired color with the "Font Color" button. I recommend choosing a tone darker than the graph line. This way the user will immediately understand which line on the chart each label belongs to.

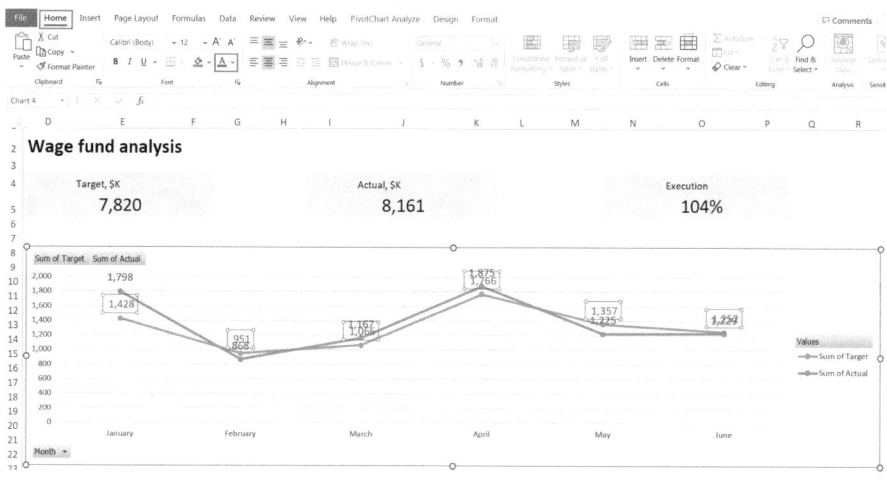

Tip!   *For the X-axis title and labels, it is better to leave the default color or choose a neutral one (for example, dark gray) – these elements should not distract from the main information.*

The Axis and Grid Are No Longer Needed

We've labeled all the data points right on the graph, so we won't need the scale on the left anymore. And since it does not perform any functions, we will delete it in order to clog up the review with unnecessary graphic elements. It is from such trifles that a "pleasant to the eye" visualization is formed.

Solution 1. Select the values of the Y axis on the graph, open the context menu with the right mouse button ➤ the "Delete" item.

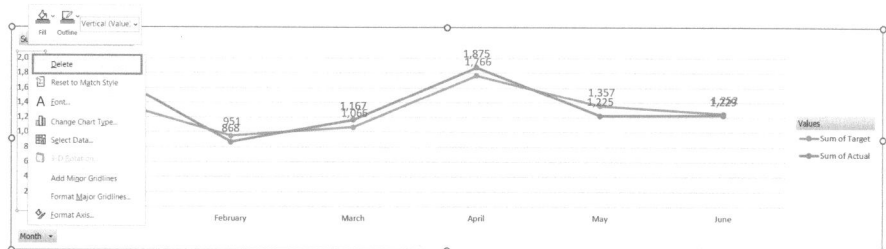

Solution 2. Select the diagram and click on the green cross that appears on the right. In the "Axes" section, uncheck the box next to the "Primary vertical" item.

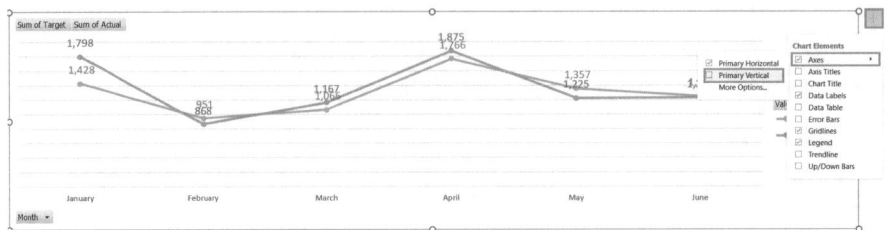

By the same principle, we will remove the horizontal grid lines on the diagram. They are needed to compare points on the graph with the scale. But if there is no scale, then these lines do not perform useful functions.

■ *Tip!   Sometimes they object to me: it looks empty without lines, let's add at least a background fill of the diagram. My answer is no. We work with minimalistic business graphics, and let's maintain this style to the end. When we finish the dashboard, you will make sure that decorations in the form of lines or backgrounds are not needed.*

To delete the grid lines in the diagram, click on any of them, make sure that the entire grid is selected, and click "Delete." Alternatively, use the "+" icon to the right of the highlighted chart and uncheck the "Gridlines" item in the drop-down menu.

Now there is almost no excess left on the chart.

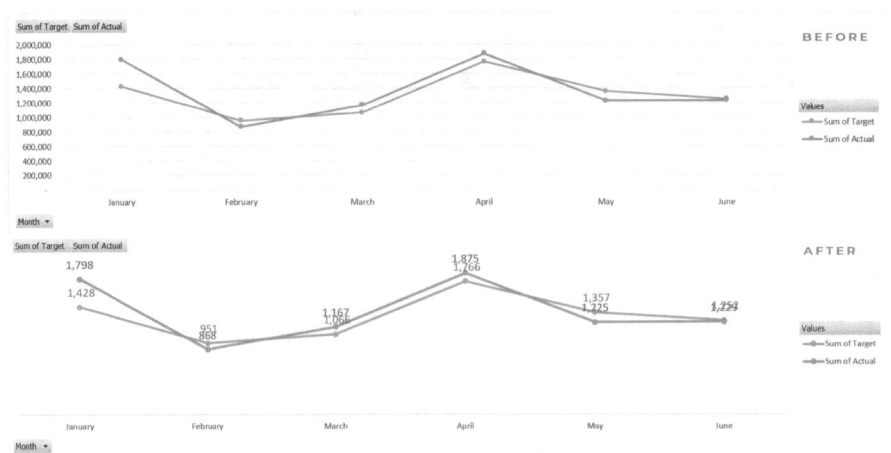

Summary

The task of labels on the dashboard is to facilitate the perception of data and make visualizations understandable. The default settings do not help this, so we brought the labels to the desired form ourselves.

- Added data labels to the "Chart Elements" menu.

- Set the decimal places number of the labels to zero on the "Data Labels format" panel.

- The decimal places were removed so that it was easy to perceive numbers.

- Increased the font of data labels, X-axis labels, and legends.

- We removed the Y axis along with the values on it: we already have data labels.

- Removed the grid lines on the diagram to avoid "information noise."

3.3 Working with the Text: Remove the Superfluous, Add the Necessary

In the last part, we simplified the perception of numbers on the diagram. Now let's deal with words and titles so that they convey the meaning of visual elements as quickly and clearly as possible.

To do this, we will work with the legend and remove everything from the graph that does not help data analysis.

Where and How to Place the Legend

By default, Excel has positioned the legend to the right of the chart. But the person starts reading from the left. That is, first he will see the graph and only then, having stumbled upon the legend, he will understand what his lines mean. So, we will have to go back to the graph again.

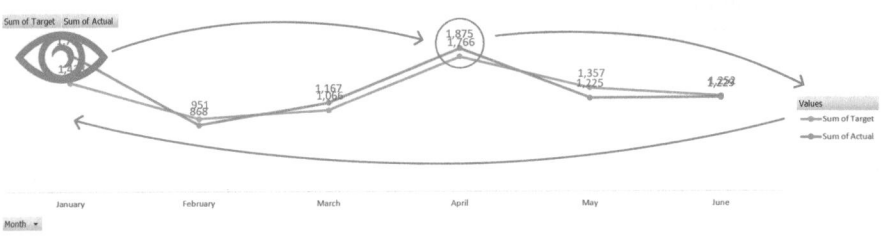

That's why the legend should be visible immediately. To do this, it should be placed above the diagram or to the left of it. That is, in the direction of reading: from top to bottom and from left to right. There are several ways to configure the correct location.

Solution 1, Through the Context Menu

We highlight the legend on the graph.

- Open the context menu with the right mouse button and select the item "Format Legend…"

- On the panel that opens, in the "Legend Position" section, select the "Top" item.

- Check the box "Show the legend without overlapping the chart."

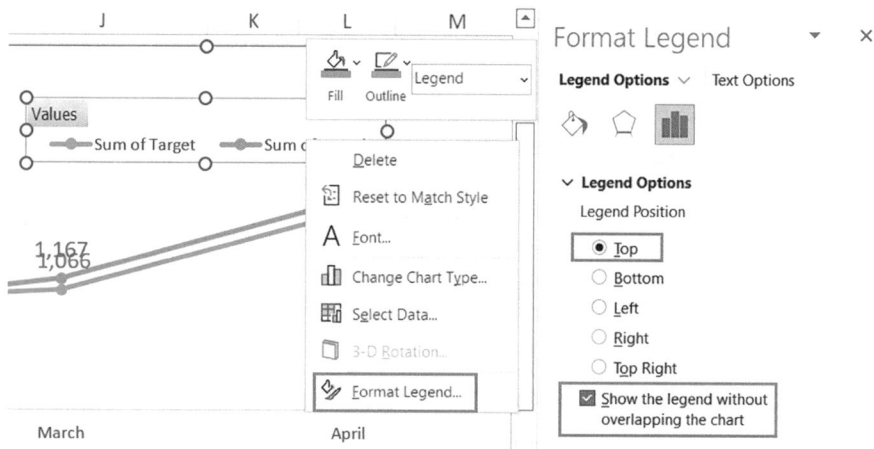

Solution 2, in the "Chart Elements" Menu

We highlight the legend on the graph.

- By clicking on the icon with a green cross to the right of the chart, we open the "Chart Elements" menu.

- In the drop-down list next to the "Legend" field, select the "Top" item.

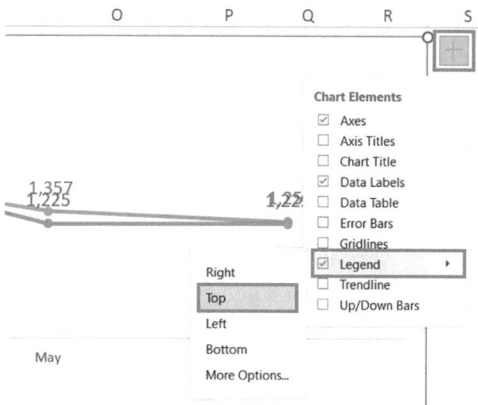

This arrangement of the legend takes into account the peculiarities of human perception and saves space on the dashboard.

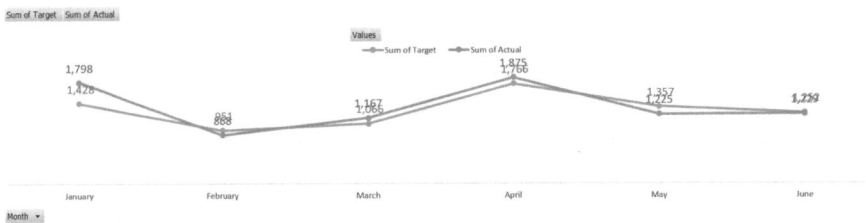

What to Do with Automatic Labels

The default location of the legend is not the only problem. It also has automatic field titles with the captions "Sum of Target" and "Sum of Actual." They need to be changed to a simple and understandable "Target" and "Actual," respectively.

There is no "Disable sum of field" button in the chart settings itself, so we will change the names of these fields in the original selections on the "Draft" sheet. There are several ways to do this.

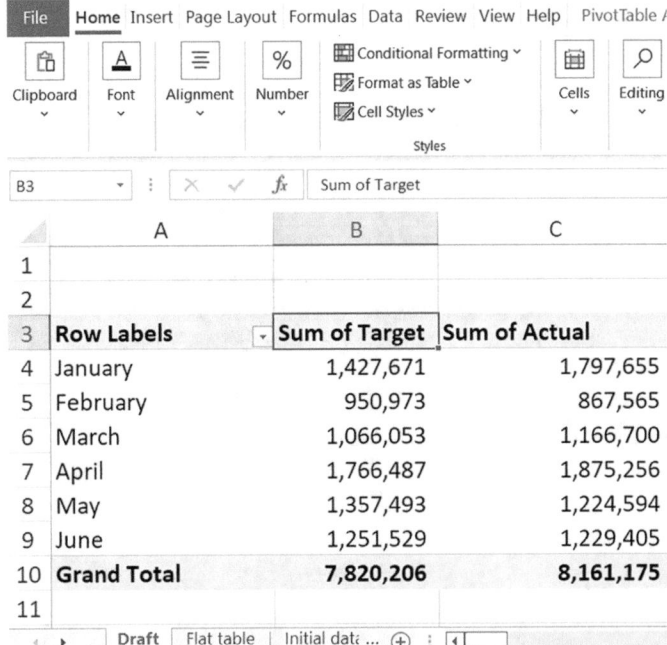

Solution 1. We put the cursor in the cell with the name of the field and remove the unnecessary inscription "Sum of." If we have several samples, it won't take much time.

Solution 2. Double-click on the name of the field in the sample – the dialog box "Value Field Settings" opens. In the "Custom name" section, delete the extra words. In the same way, you need to edit the name of each field.

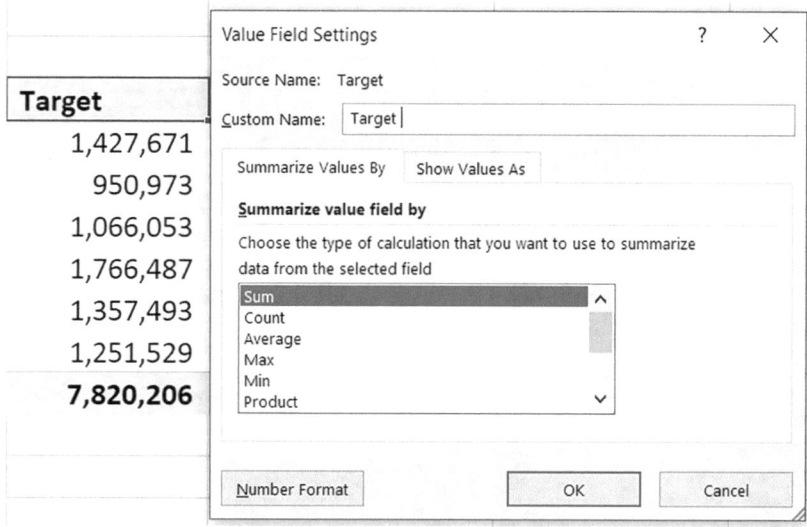

When editing the field name, leave a space at the beginning or at the end to avoid an error:

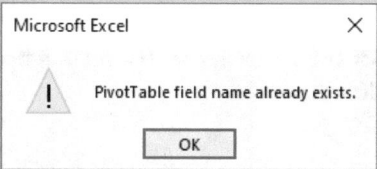

For visualization, this space does not play a role, but for the program there will be a difference with the name of the field in the pivot table.

Solution 3, recommended. The text "Sum of" can also be removed using autocorrect – this is the fastest option.

Being on the "Draft" sheet, on the "Home" tab in the "Editing" section, we find the "Find & Select" button. Select "Replace" from the drop-down list.

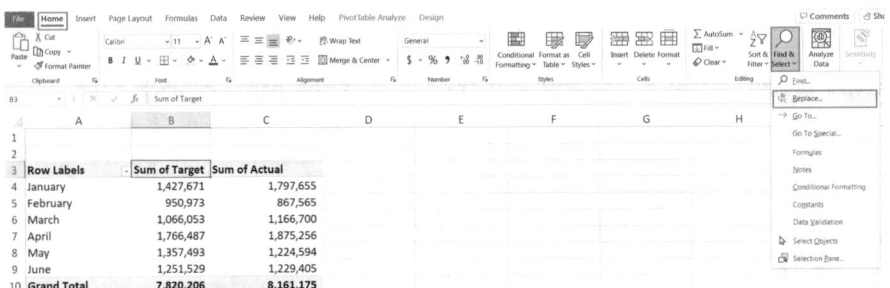

A dialog box appears – it can also be called with the keyboard shortcut Ctrl + H (or cmd + H). On the "Replace" tab, we have two fields. In the first one, enter the words "Sum of," and leave the second one empty and click "Replace all." A window will appear with a message about the execution of the command and the number of changes made.

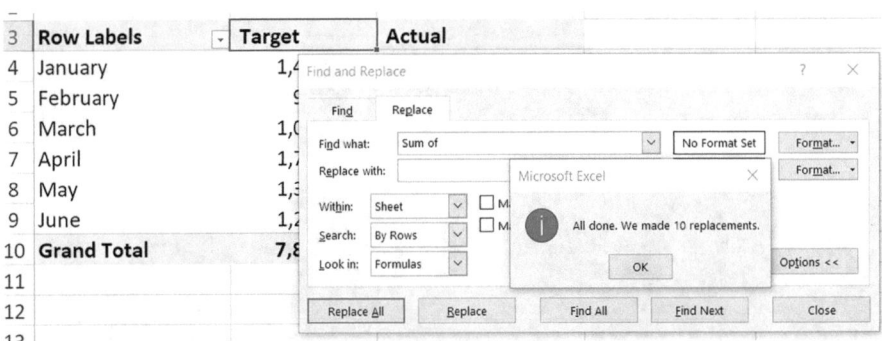

After that, the names of the fields in the legend will be updated automatically and become more understandable.

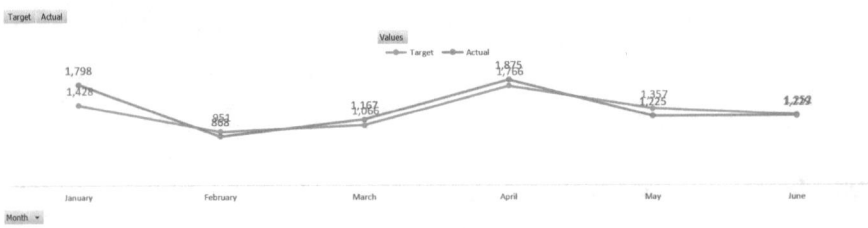

How to Delete Field Buttons

We built a chart based on a pivot table, so Excel automatically added several buttons: on the chart, these are "Target," "Actual," "Values," and "Month." They don't perform any useful functions on our dashboard. You can only click the "Month" button and open the filter, but it is more convenient to use a slicer for this. The buttons take away almost half of the working area of the diagram, so we will remove them.

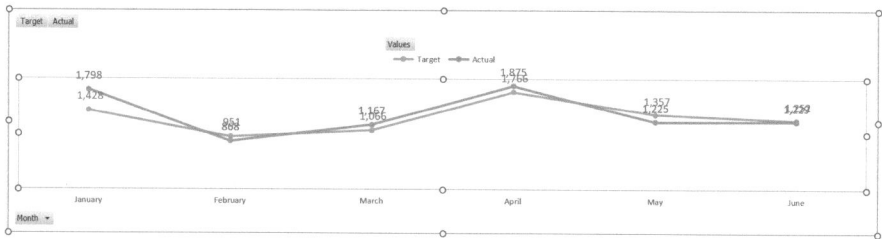

Solution 1. Warning: this method does not work in all versions of Excel. Select the button on the diagram and open the context menu with the right mouse button. Select "Hide All Field Buttons On Chart."

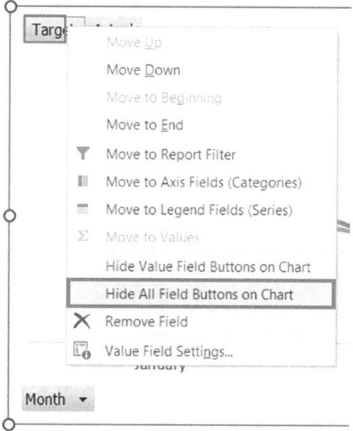

■ *Tip!*   *Never select the "Remove field" option, otherwise the program will delete the original field with the data in the pivot table.*

Solution 2. Select the chart and on the menu tab "Pivot Chart Analyse" (in older versions of Excel this is the "Analyze" tab) you will see "Field buttons" on the right. In the drop-down menu under this button, check the "Hide all" item.

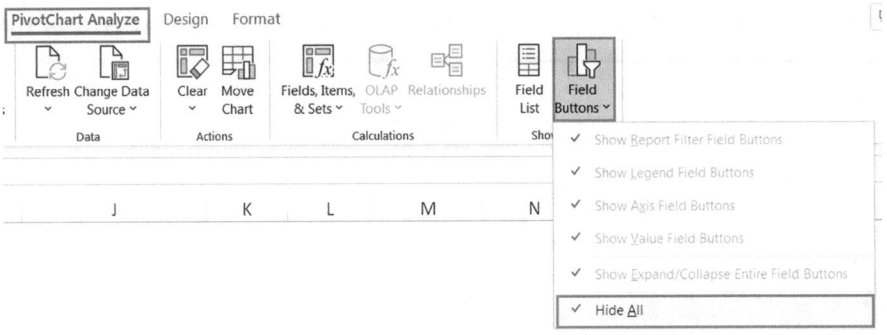

That's it – unnecessary elements are removed, and there is twice as much space for the graph itself!

Two Ways to Add a Chart Name

In order for the user to immediately understand what the diagram shows, you need to add a meaningful title to it. If we were doing a visualization for a slide, then we would try to formulate a conclusion in the title right away. But we have a dynamic dashboard, and the content of the chart depends on filters. Therefore, we will briefly indicate the keywords in the title.

Solution I. On the "Design" tab, click on the "Add Chart Element" button on the left. In the drop-down list, select "Chart Title" ➤ "Above Chart." A field will appear on the chart in which you need to enter the name of our schedule – "Payments dynamics."

Solution 2. Click on the cross to the right of the selected chart and tick the box "Chart Title" – as in the first method, a header field will appear.

Placing the title in the center with an overlay would help to increase the working area of the graph. But this option is only suitable for static slides if the line or column does not overlap the title. On the dynamic dashboard, we select the option "Above the chart" so that such an overlap does not occur when changing or filtering data.

It remains to set the font size for the chart title: it should be larger than the labels and legend: from 14 to 18, depending on the free space. To do this, select the block with the title and set the desired value on the main menu tab in the Font section – I chose size 16.

Summary

The diagram should have a minimum of text – only what helps the user to easily read and understand the information. To solve this problem, we have performed several actions:

- We have set up the location of the legend above the diagram so that the content of the graph is immediately clear.

- Removed the "automatic" labels from the legend by changing the names of the fields in the pivot table.

- The useless field buttons inherited from the pivot table were removed from the chart area.

- Gave the graph a clear title and placed it on top.

- Increased the font size for the chart title to 16.

3.4 Designing Bar Charts

Data labels, format and bit depth, removal of the axis and gridlines on the diagram, legend, extra buttons, name – all this is relevant for setting up any visual element, not just for the graph.

But in the charts on our dashboard, there is an element that is not on the graph – these are columns. Formally, "Expenses by divisions" and "Expenses by items" are different types of visualization

- With vertical columns – column chart

- With horizontal columns – a bar chart

Both can be called bar charts. They have the same setup principles, except for one nuance.

To make it easier to perceive and compare information, we will adjust the distance between the columns and set their sorting so that the best and worst indicators are visible immediately.

In this part, I will illustrate the actions on already configured diagrams so that unnecessary elements do not distract your attention. You will add these settings later, using a special checklist at the end of the part.

Distance Between Columns

By default, an emptiness has formed between the columns with the target and the actual. We also need these indicators to look like a single element for each division. To do this, "glue" the columns.

By clicking the mouse, select any row of columns ➤ open the context menu ➤ select the item "Format Data Series …"

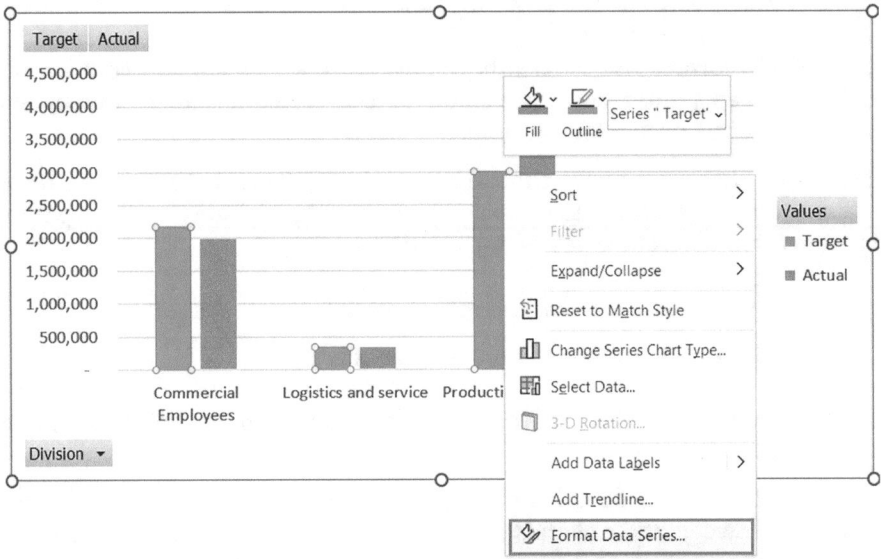

In the panel that opens on the right, in the "Series Options" section, we configure the values:

- "Series overlap" – 0%
- "Gap width" – 50%

Now the columns with target and actual values for each division look like single elements, and this helps to focus on the difference between them.

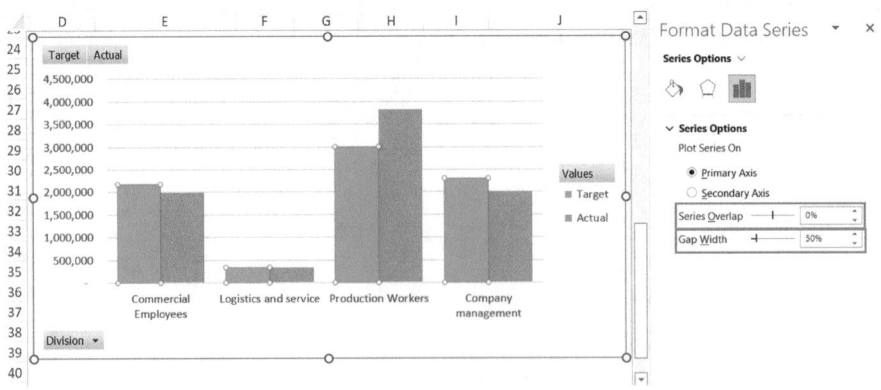

Sorting Columns in Descending Order

It is always easier to perceive ordered values. So that the user's eyes do not rush between columns of different heights, but consistently go from larger to smaller, we will set up sorting.

The order of columns in the chart depends on the sequence of rows in the pivot table. The dynamics chart is built according to the name of the month – they were immediately ordered correctly. And in the samples for the other two visual elements, you need to adjust this sorting in the "Actual" field from a larger number to a smaller one.

There are two ways to sort columns on a column chart.

Solution 1: on the diagram. We select a column in the diagram "Expenses by division." Since we want to compare divisions primarily by actual results, we select "Actual" columns. Right-click on the context menu and select "Sort" ➤ "Sort Largest to Smallest" in it.

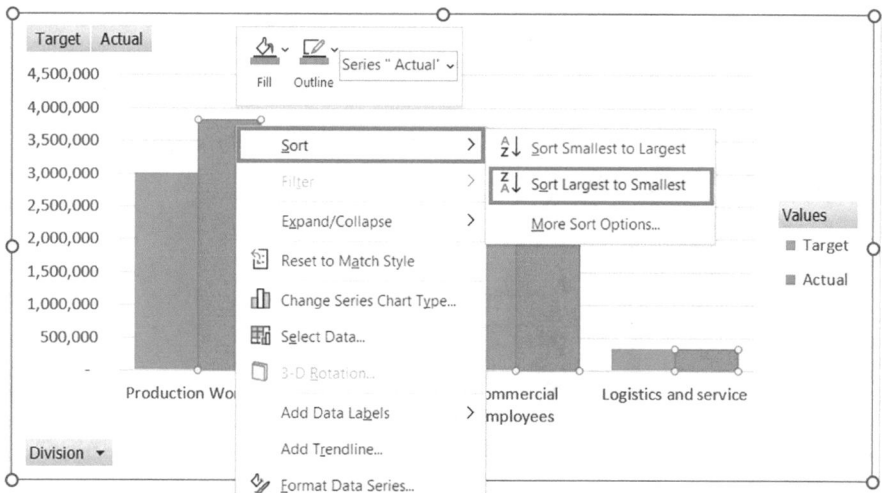

Solution 2: in the pivot table. On the "Draft" sheet, put the cursor in the cell of the "Expenses by division" sample. Right-click on the context menu, select "Sort" ➤ "Sort Largest to Smallest."

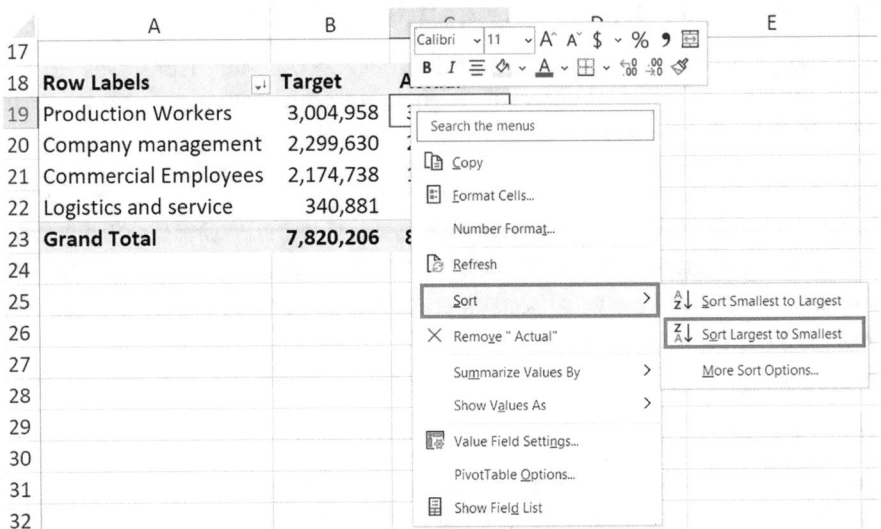

The order of the chart columns on the "Dashboard" sheet will change automatically and will take a logical form for human perception.

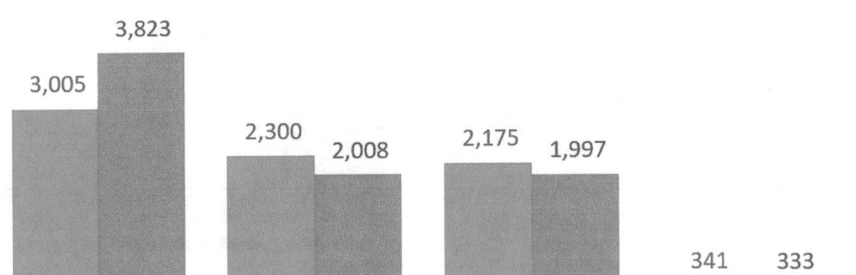

___

■ *Tip!*   *I sort data, adjust the format and bit depth of the numbers immediately after creating the pivot table. This helps not to come back to this later when designing the diagram.*
___

Sorting Columns on a Bar Chart

With a bar chart, everything is not so simple: Excel builds data categories in reverse order by default.

Even if you set the table to sort "In descending order," as we did with the column chart, the categories will line up exactly the opposite on the chart.

Expenses by items

■ Actual ■ Target

Civil contracts	142 / 123
Vacation reserve	470 / 485
Extra-budgetary funds	1,517 / 1,343
Bonus	2,077 / 1,682
Salary	3,955 / 4,187

To fix this, you need to set up reverse sorting in the data sample – "Sort smallest to Largest." Then (oh what a miracle!) on the bar chart, they will rise in descending order. I will not dwell on the reasons for such a construction, just accept it – Excel is such an Excel...

Row Labels ⤵	Target	Actual
Civil contracts	123,447	142,270
Vacation reserve	484,658	470,414
Extra-budgetary funds	1,343,341	1,516,563
Bonus	1,681,995	2,076,739
Salary	4,186,764	3,955,189
Grand Total	**7,820,206**	**8,161,175**

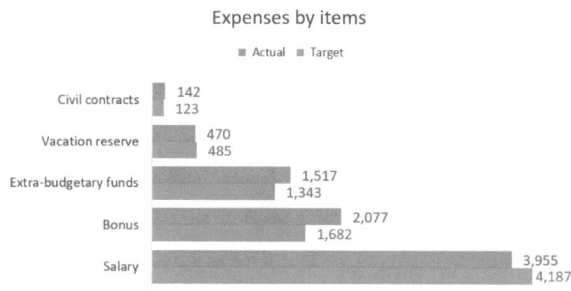

The same thing happens with the order of the "Target" and "Actual" columns. In the diagram, both in the legend and in the columns, the "Target" comes first. And only in the bar chart, the "Actual" comes first. Let's fix this by adjusting the axis format.

Select the categories on the Y-axis with the right mouse button and select "Axis Format" in the context menu.

In the panel that appears, in the "Format Axis…" section, specify the following:

- In the "Horizontal axis crosses" group, select "At maximum category."

- In the "Display units" group, check the "Categories in reverse order."

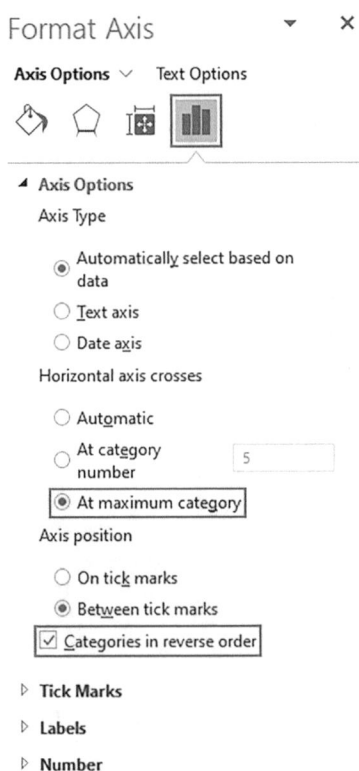

These parameters will automatically arrange the rows according to the sequence of fields in the sample and we will not have to do anything extra.

I recommend that you always choose the second method to configure the bar chart.

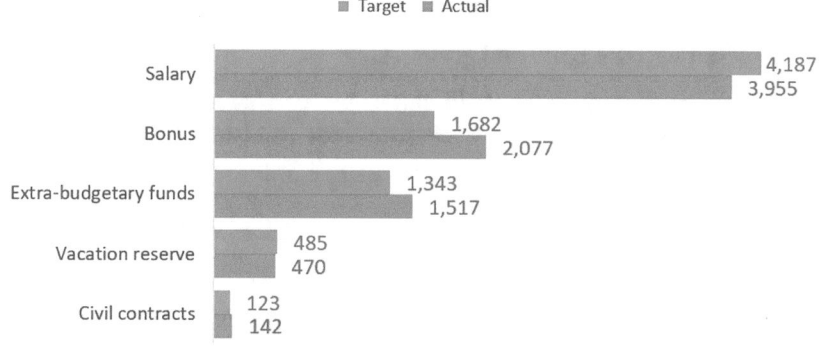

Chart Design Checklist

Let's work out the design peculiarities on the "Expenses by division" chart: configure this visual element with a short checklist.

1. Add labels of data, select 12 pt font for them.

2. Change the bit depth of the numbers to thousand or million, so that there are 3–4 characters left in the labels.

3. Remove the scale and gridlines.

4. Color the data labels in the color of the row (a tone darker), do not add a background for the labels.

5. Hide the buttons of the pivot table fields.

6. Place the legend on top (sometimes on the left).

7. Remove automatic labels from the legend.

8. Add the title of the chart, select 16 pt font for it.

9. Reduce the gap width between the chart columns by up to 50%.

10. Adjust the sorting of columns from larger to smaller.

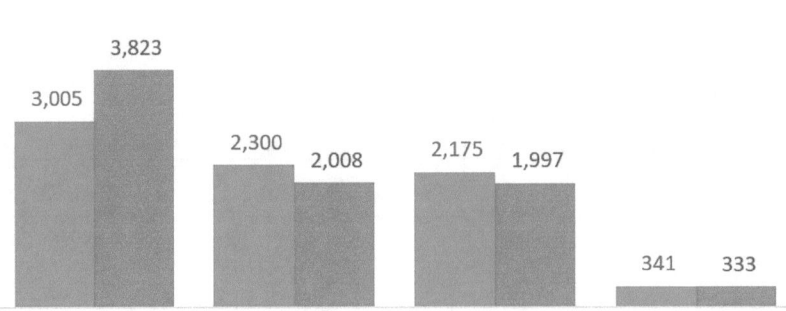

Expenses by division

■ Target ■ Actual

3.5 Setting Up the Chart Template

Working on each diagram is monotonous – not everyone likes it and it takes time. This process can also be optimized if you once set up a diagram and save it as a template, which will be useful for other work.

We will make a template from the "Expenses by division" chart, which we have already set up according to our checklist.

How to Save a Template

Everything is simple here: right-click on the block with the diagram "Expenses by division" and select "Save as Template..." in the context menu. You can select both the entire diagram and only the plot area – only the length of the context menu will differ, but the necessary item is available in both cases.

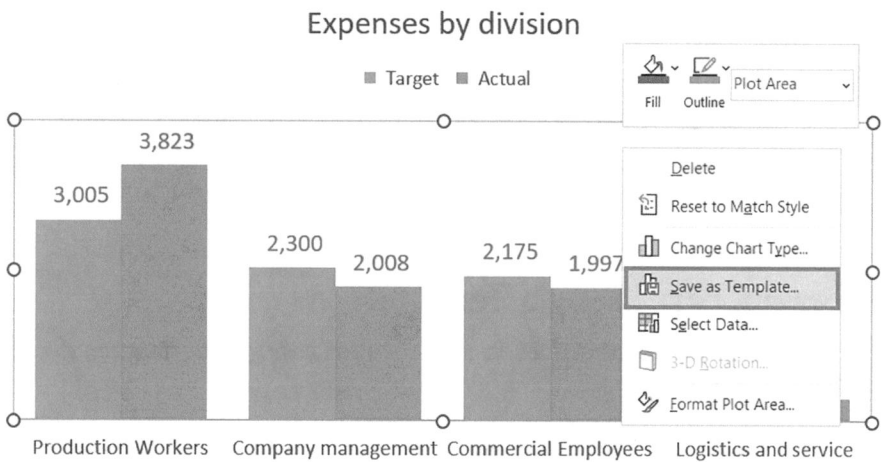

▉ *Tip!*   *Never change the path to the folder to save. Excel automatically loads custom templates only from it.*

Do not forget to give the templates clear names when saving, so that you can easily find the desired diagram in the list later. Our template is saved with the name "Chart1."

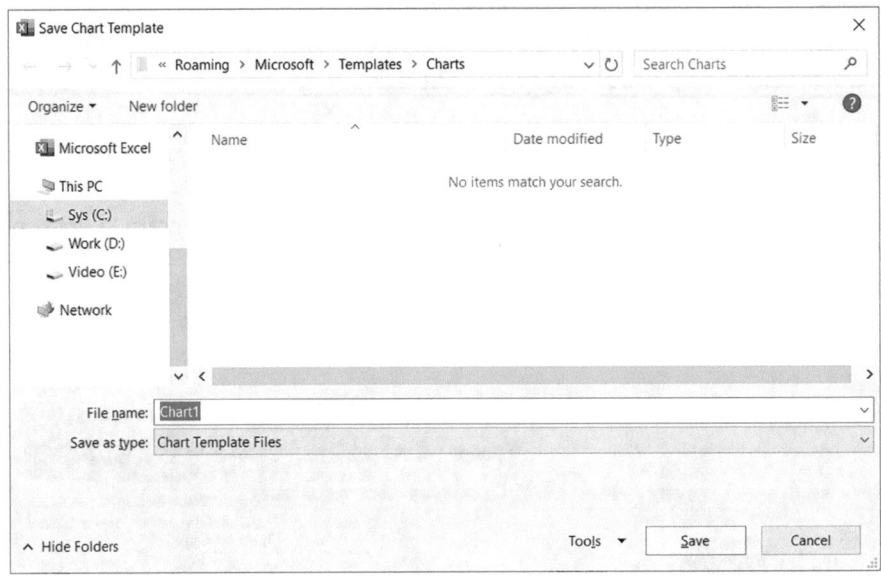

A New Chart Using a Template

To apply a ready-made template to another (not configured) diagram, do the following:

- Select the entire diagram or the plot area on it.
- In the context menu (right-click), select "Change chart type."

A window opens with a menu on the left – go to the "Templates" folder and select our saved template "Chart1."

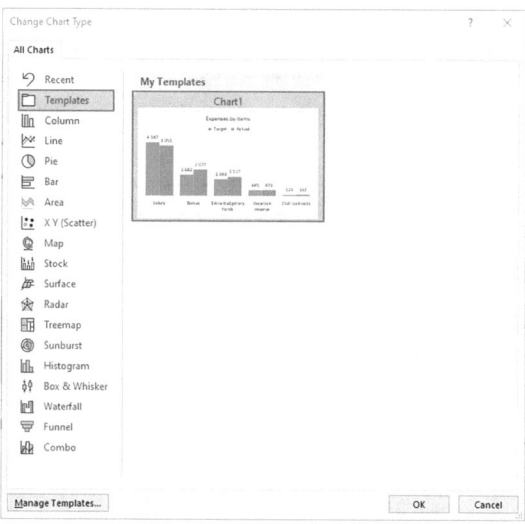

All the settings of our template are automatically applied to the selected diagram: the location and color of the labels, the bit depth of the numbers, and everything else. It remains to change the name to "Expenses by items."

Expenses by divisions, $K

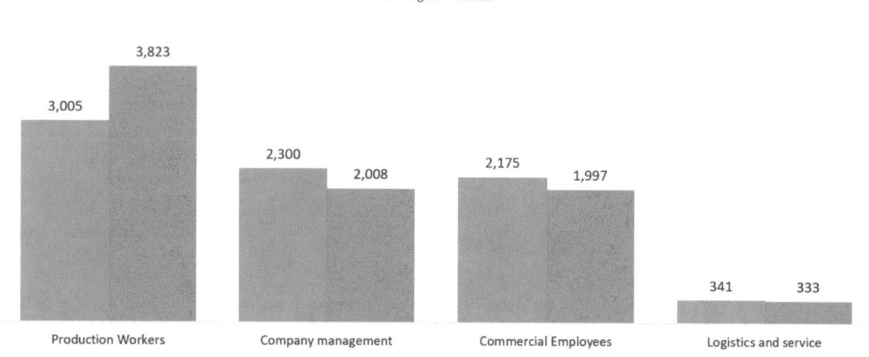

Adapting the Template to the Diagram

We need to visualize expenses by items in the form of a bar chart with horizontal columns. But when using the template, the appearance of the visual element has also changed – the columns have become vertical. Let's fix this by re-selecting the chart type.

Select the entire diagram or only the plot area.

1. Open the context menu and select "Change Chart Type."
2. In the window that opens, select the "Bar" chart type.

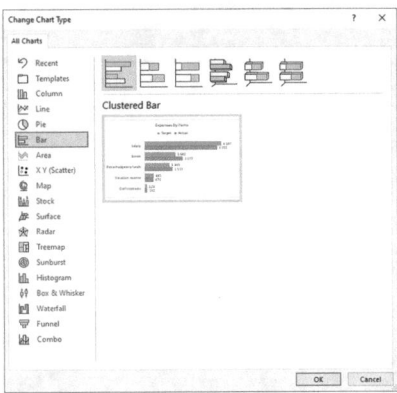

After completing these steps, the "Expenses by items" chart takes the desired form in accordance with the layout.

It remains to change the sorting and the order of columns in the bar chart. That's it – we've put together a full-fledged draft of the dashboard!

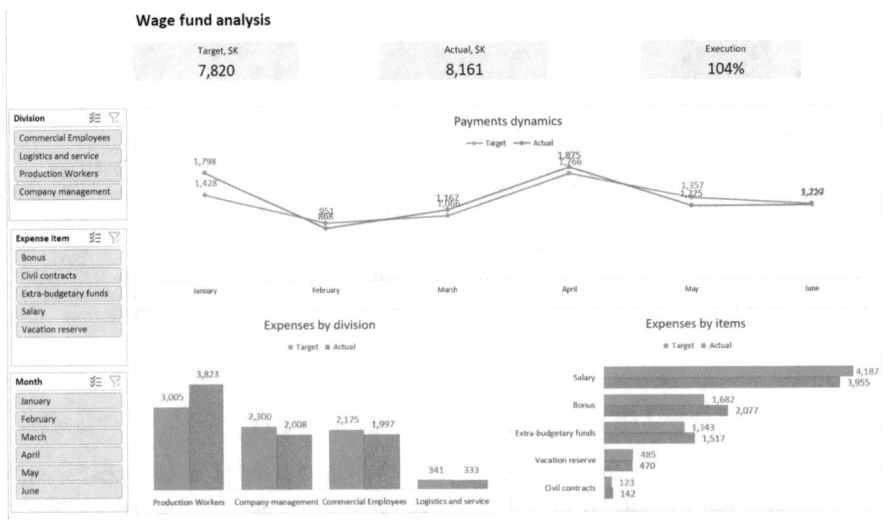

Summary

The design of each visual element takes time. And since there are several of these elements on any dashboard, it will take a lot of time. Using a template will help simplify and speed up the process.

- After setting up the diagram, we save it as a template without changing the path to the folder.

- Select a new "draft" chart and use the "Change chart type" option to apply the saved template to it.

Adapt the template by re-changing the type of visual element to the one that is needed in the new diagram.

Final Design

We now have a dashboard in a minimalist style, which follows all the rules of data visualization. It is still designed in standard Excel, while customers usually prefer a variety of designs.

To create a new style, we will go beyond traditional chart settings and use a couple of life hacks. The standard dashboard will then look like a designer model.

I will keep working in the minimalist style chosen. It is still actual: in the context of information overload, it is vital to remove everything superfluous and preserve only but significant shapes, numbers, and text labels. Information design is not ornamentation.

The techniques we use are also suitable for other data visualization tools, Russian BI systems included. The only difference is configuration. It is here that we employ original life hacks to work in MS Excel.

© Alex Kolokolov 2023
A. Kolokolov, *Make Your Data Speak*,
https://doi.org/10.1007/978-1-4842-8942-6_4

4.1　Aligning the Headers to the Grid of Cells

It is not always possible to align all dashboard elements by eye. If you look closely, you will spot inaccuracies in my example, which is very difficult for a beginner: fitting "pixel to pixel" requires both effort and time.

Standard chart titles present a similar problem. You cannot change the size of this block: if the header is long, it will take two lines. It will then move the legend down and break the symmetry between the neighboring diagrams.

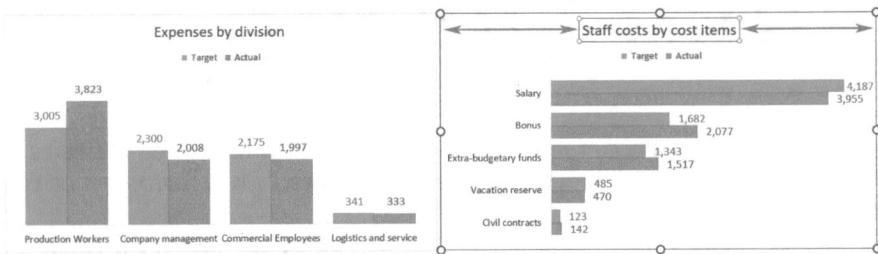

To avoid disharmony, I recommend placing headers in cells rather than in standard "Chart Title" elements. They will still be in Excel even if the grid lines are disabled, which solves the alignment problem.

■ *Tip!　I recommend making a copy of "Dashboard" sheet for all further actions. This way you will save the draft version and see the difference between a standard dashboard in Excel and a professional one.*

Adding a Headers to the Cells

Working with the new dashboard, we will first create and arrange headers in the cells and then disable the standard titles on the diagrams.

Move the top graph slightly lower to clear the header row. Enter the graph title "Payments Dynamics" in the cell above the left corner of the graph.

Applying Cell Style

We now need a style for the header that will be visually different from a usual text cell.

Find "Cell Styles" section ("Styles" in older versions of Excel) in "Home" menu tab and select "Header 2" from the drop-down menu. This option will add a light blue outline to the lower border of the cell with the header.

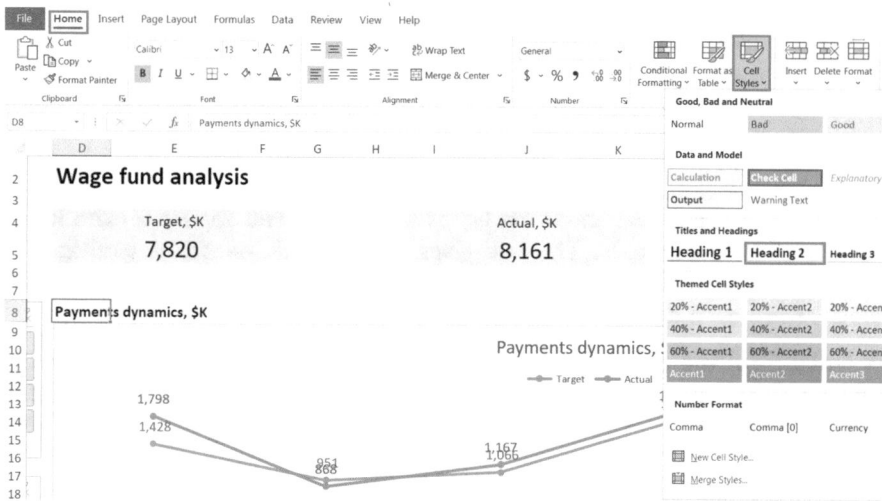

This outline will visually separate semantic blocks on the dashboard from each other. It will also make a slight emphasis on the symmetry and accuracy of the elements on the dashboard.

Excel default font size for new headers is 13 points – we will set it to 16. We will then disable the duplicate standard title from the diagram block using the green cross to the right of the diagram ("Chart title" item) or in "Design" tab ("Add Chart Element" button).

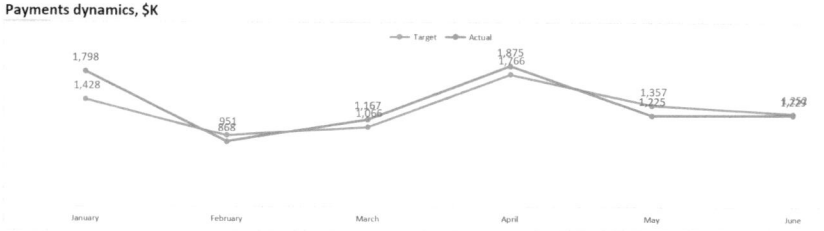

Copying Formatting to Adjacent Cells

We will use the same format for several more cells. Put the cursor in the cell with the chart title ➤ click the "Format painter" button in "Home" menu tab ➤ select the cells after the header to the end of the graph with the mouse.

If everything is correct, the same light blue outline will appear on the lower border of all these cells. The line gives an aesthetic effect, as well as sets boundaries for all elements on the dashboard and visually separates semantic blocks from each other.

Now turn off the outline of the diagrams – this will save our product from typified Excel features and give extra airiness to the report. To do this, you need to

- Select the block with the chart.

- Click "Shape Outline" in "Format" menu tab.

- Select "No Outline" in the drop-down list.

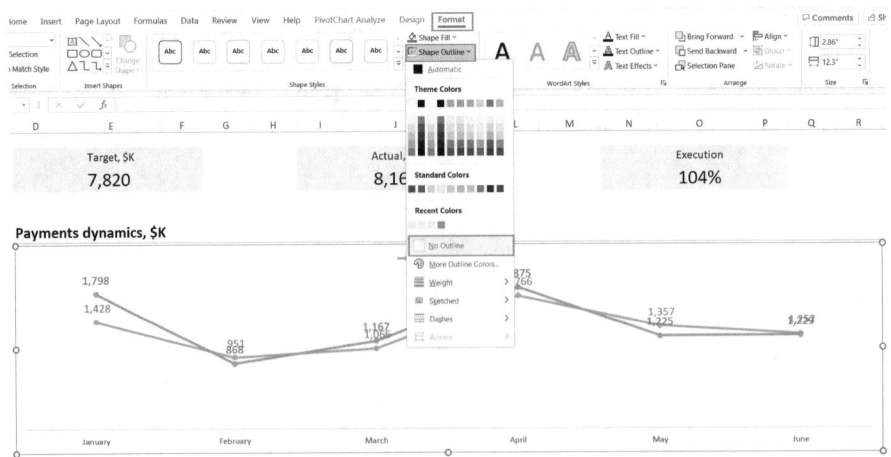

Without the contour, we now have more free space. Raise the legend a little higher and expand the area of the diagram. This will draw attention to the data.

Tip!   *Align the titles to the left rather than the center. This will help avoid leaving an empty space on the dashboard. You will be able to raise the legend higher without piling up elements on the dashboard.*

At this stage, there are enough lines under the new headers to create a clear dashboard structure. They are also tied to the Excel grid and will not be displaced in further visualization work. This is what the chart titles will look like.

Summary

We started a new design version of the dashboard by designing chart titles in cells. They divided the whole of the panel into semantic blocks and set a clear structure for it.

We added headers to the cells above the upper left corner of each chart.

We aligned them to the left to avoid leaving an empty space.

We chose "Header 2" style for cells with headers.

1. We increased the font size to 16 to define semantic blocks.

2. We disabled standard names inside the charts.

3. We copied the style for the cells to the right of the headers.

4. We disabled the diagram contours and expanded the plotting areas.

4.2 Creating New Cards on Top of the Cells

The chart titles are now in the cells, which is not immediately noticeable. This simplifies the alignment and increases the working area of the diagrams. There are side effects, too.

The distance between the lower diagrams equals the width of the cell, which is sizeable by itself. To shorten this distance, you will have to reduce the width of column J. This will affect the card with the actual indicator: the value will no longer fit the column and will be displayed as several ### characters.

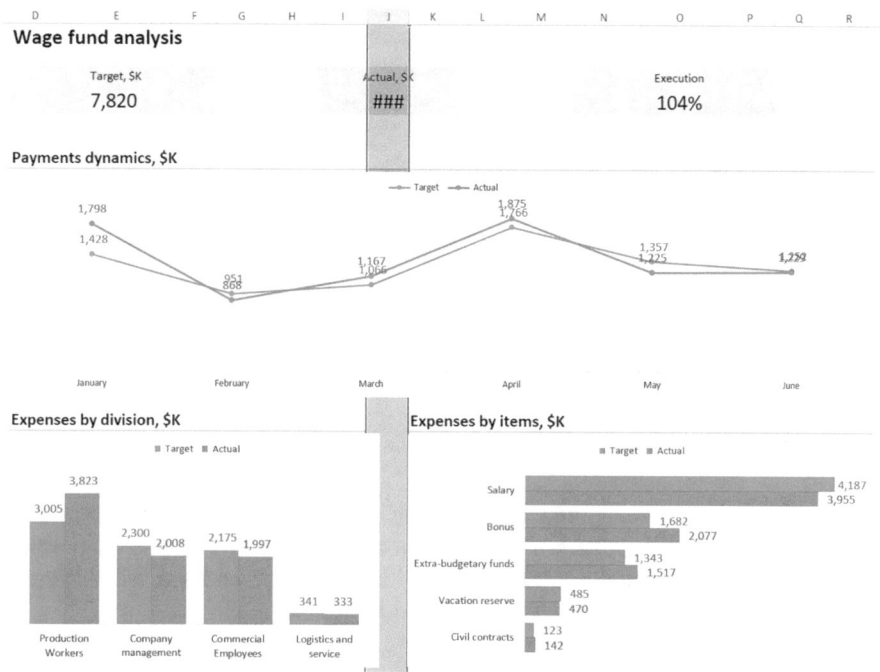

Combining cells seems to be a workable solution to the problem. Still, this will always come up if the cards "live" inside the cells.

For example, I want to add another card with average monthly salary. I will then have to compress and move the other cards, which in turn will affect the overall diagram alignment.

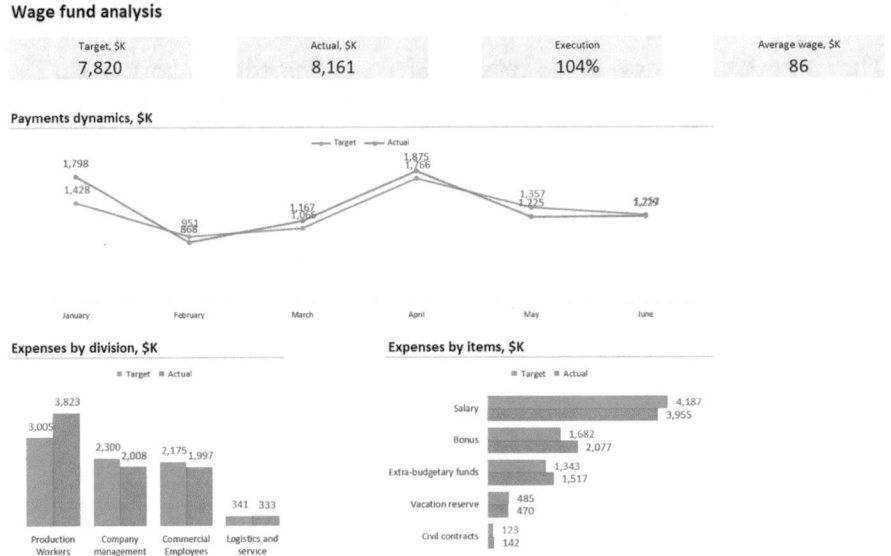

When creating a dashboard and negotiating it with the customer, you may have to add and remove cards and diagrams repeatedly. To avoid ruining the alignment, we will "free" the cards from their grid cell bounds. We will create new independent elements.

How to Add a Card Using a Shape and Text Box

Modern business intelligence systems have widgets for cards. You add an indicator, and there appears a stylized rectangle with a title and final value. There are no ready-made objects of this kind in Excel, and we will assemble one using makeshift tools – rectangles and text boxes.

It may seem ridiculous, but this life hack will be helpful in Excel as much as in other programs. It comes especially handy when the dashboard has a non-standard layout, and several indicators need to be grouped under one heading. Let us take a closer look at the steps of the process.

Step 1. Click the "Shapes" button in "Insert" tab in "Illustrations" section ➤ select "Rectangle" shape in the drop-down menu. If you have a large screen, "Shapes" button will immediately appear on the ribbon.

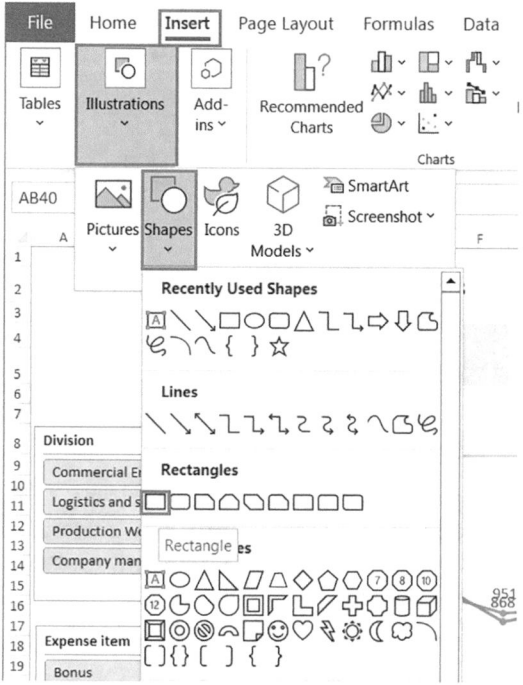

Step 2. Place a rectangle on top of the first KPI card. It can be set to any size by pulling the edges of the shape itself. This will not affect the width of the columns in the table.

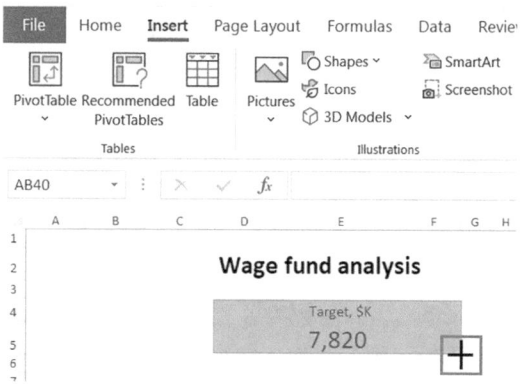

Step 3. Select the rectangle and type in the title of the card: "Target, $K".
Let's set the font size (12) and center alignment.

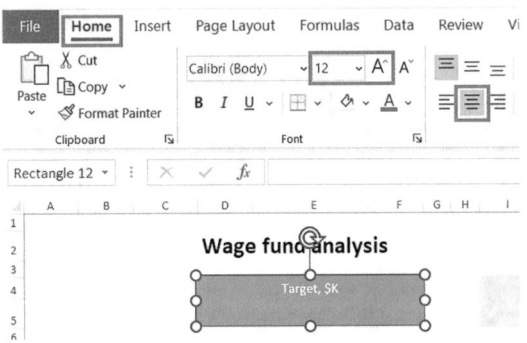

Step 4. Click an empty cell to de-select the rectangle. Click the "Text box"
button in "Insert" tab in "Text" section and place the item appearing into the
rectangle. This is the element where we will add our indicator.

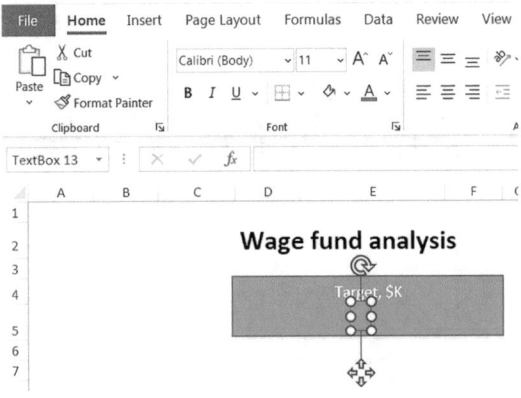

To add an indicator to the card, we need to create a new cell in the pivot
table sample.

Totals for Cards in Separate Cells

We wrote out formulas in the draft dashboard to display the card indicators in a convenient format and avoid using too many characters. You need to configure this format too, with account for its specifics.

The formula cannot be specified in the formula bar for "Rectangle" and "Label" elements. If we try and do this, we will see an error message.

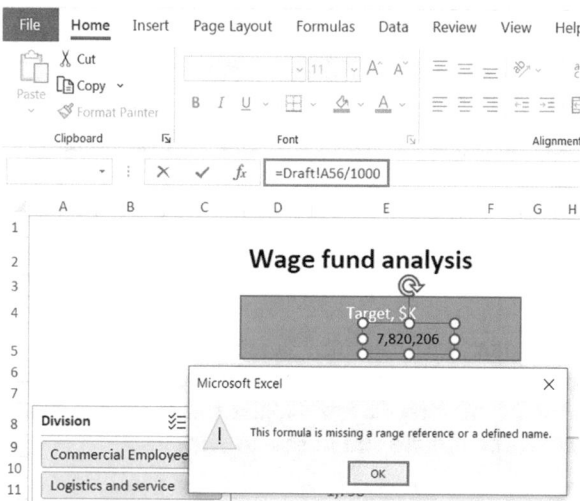

To add a numeric indicator to a new card type, we need to create cells with formulas under "Key indicators" sample in "Draft" sheet.

Step 1. Select an empty cell under the first column of the sample and enter the following values into the formula bar: = A43/1000, A43 being the address of the cell with the plan indicator where you need to reduce the number of characters.

Step 2. Set formatting for the new cell as before. This reduces the bit depth and leaves an integer without decimals.

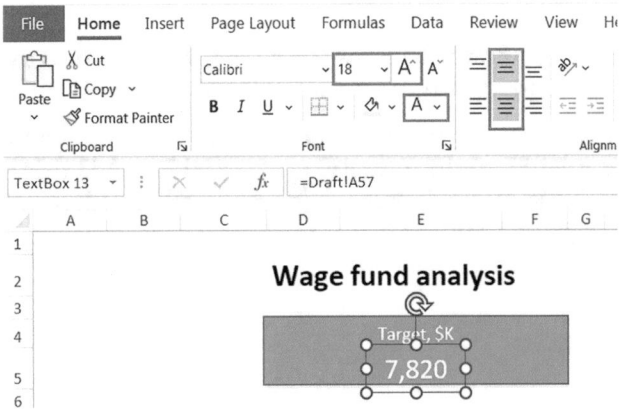

Step 3. On the sheet with the new version of the dashboard, select "Text box" element added to the card and specify a link to the new cell with the desired indicator in the formula bar. Adjust the font size, color, and position of the indicator in the middle and in the center.

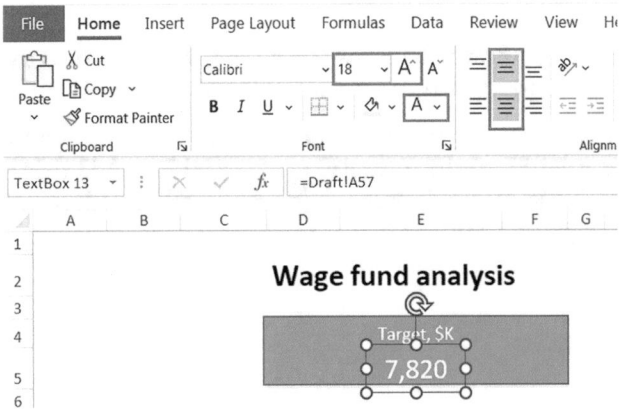

In the same way, prepare the cells with formulas for the other two cards in advance:

- Set the formula for "Actual" indicator and then reduce the bit depth by removing decimal points.

- In the formula itself, write the division of the actual indicator by the targeted value for "Execution" card. Convert the result into percent in "Home" tab in "Number" section.

Step 4. Align the card elements relative to each other. Hold down Shift key and select "Text box" element with the mouse, then select "Rectangle" element. Select "Align" ➤ "Align Center" in "Shape Format" menu tab.

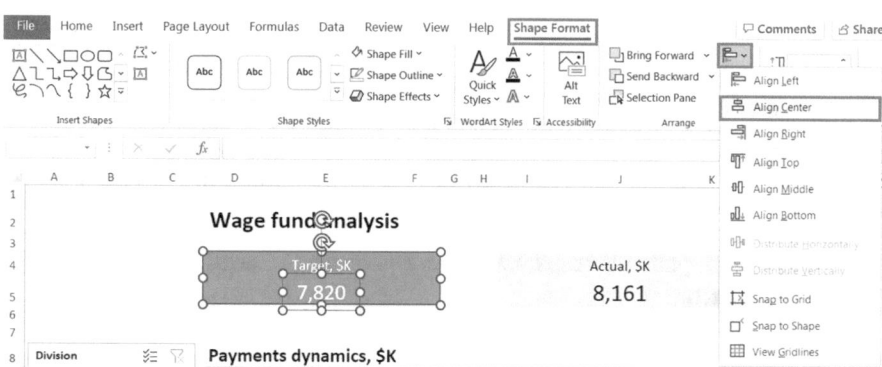

Step 5. Combine the elements of the card. Select "Text box" first, then select "Rectangle." Use one of the following grouping methods:

1. Select "Align" ➤ "Snap to shape" on "Shape Format" tab.
2. Select "Group" ➤ "Group" in the context menu.

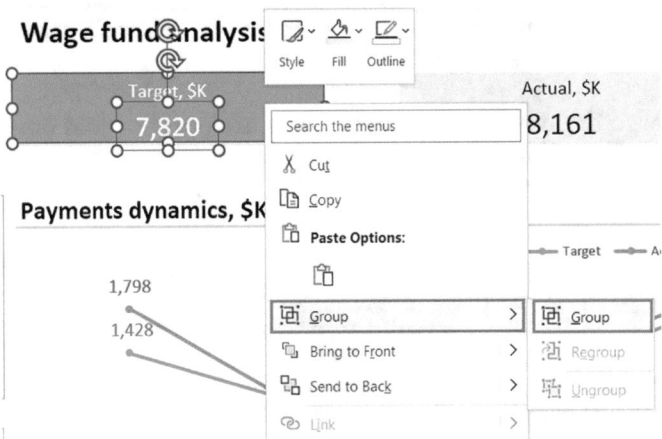

After completing all the steps, we have a grouped "Card" element with the targeted indicator.

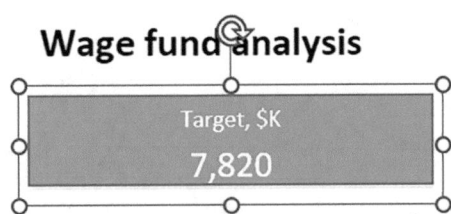

Card Replication

Our dashboard should have three cards with key indicators: "Target," "Actual," and "Execution." We will use the familiar life hack with replication to avoid setting every card separately.

To do this, I select "Target" card, hold down Ctrl (or cmd) key on the keyboard, and move the copy of the element to the desired place with the mouse. We will turn this copy into a card with the actual indicator.

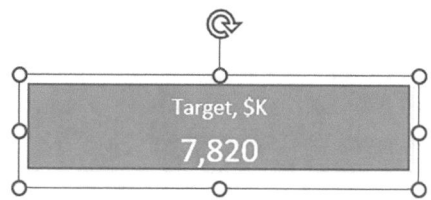

We will title the new card "Actual, $K" and change the link to the actual cal-culated indicator on the "Draft" sheet.

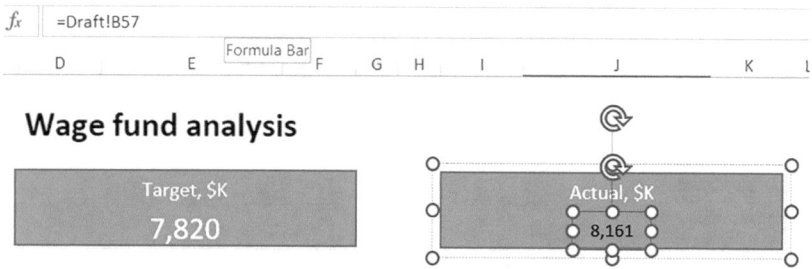

At this point, the font of the formatted label goes back to default black. It may seem that you have done something wrong. Not to worry – this is yet another Excel "bug." Take a deep breath and readjust the font manually: increase the font size and change the color to white.

How to Fix the Background Style

The main thing on the KPI cards are indicators – they are the first to meet the eye, after all. Here, however, the bright and "heavy" fill steals the limelight. We will need to change the style of the element to make the indicators contrasting and visible.

■ *Tip!*   Don't fall for originality, keep the style minimalist – a dark font against a light background (unless the dashboard design calls for something different).

Open the drop-down menu with shape styles in "Shape Format" tab ("Format" in older versions of Excel).

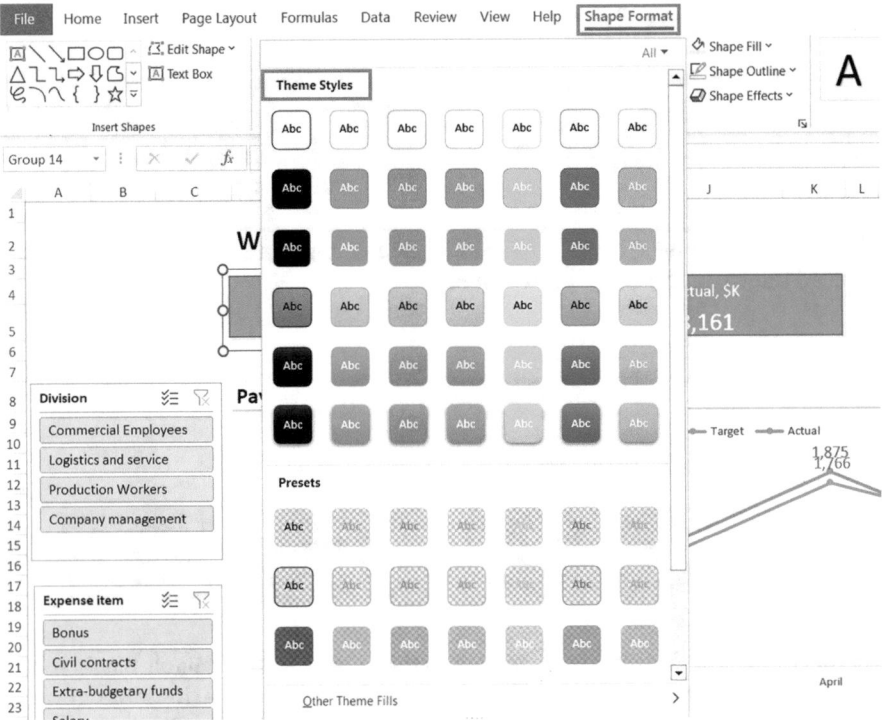

None of the options given suits our corporate dashboard design. We need to give up ready-made solutions and create something of our own. We will go to the "Shape Format" menu tab and set the following parameters:

- Shape fill: the lightest blue
- Shape outline: no contour
- Text fill: black

Now we have the minimalist style needed.

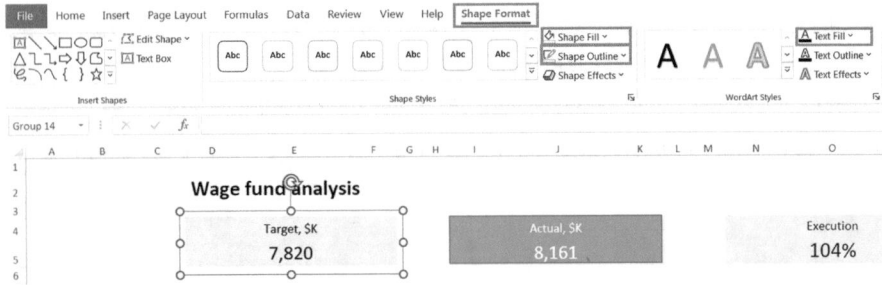

We repeat the same steps for "Actual" card and then replicate the element configured for "Execution" card, having set the formula and formatting on the "Draft" sheet in the "Key Indicators" sample.le.

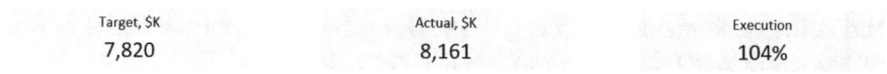

Target, $K	Actual, $K	Execution
7,820	8,161	104%

In this example, we choose a pale fill and remove the contour. You can do the opposite: remove the fill and leave a colored outline. If you do, choose the line color a shade brighter and increase the thickness to 2 pixels.

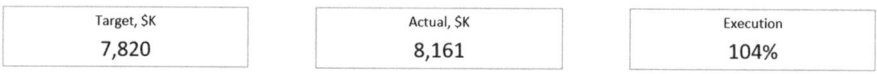

Target, $K	Actual, $K	Execution
7,820	8,161	104%

We have created cards that are not linked to the Excel grid. You can change their size to your liking: this will not deform the remaining elements on the dashboard.

Alignment of New Cards

When the cards are removed from the Excel cell level, you can quickly align them on the dashboard. It is much easier now that you no longer have to adjust the elements by eye.

Step 1. Alignment along diagram borders. First, align "Target" card along the left border of "Expenditure Dynamics" graph. Select both elements and click "Align to the left" in "Format Shape" tab under "Align" button.

The third card "Execution" needs to be aligned along the right border of "Expenditure Dynamics" graph. We will also select both elements and click "Align to the right" in the same menu item.

Step 2. Vertical alignment. It is quite simple as well: we select all the cards and in the same menu, select alignment along the upper corner, in the middle or along the lower corner – whichever is most suitable.

Step 3. Horizontal alignment. To make the cards equidistant, select all of them and click "Distribute horizontally" in the menu.

We have set up the perfect location of the cards on the dashboard in three simple steps. The cards are now aligned both relative to each other and the other elements.

Summary

KPI cards that are "independent" of the Excel grid give more opportunities for design. Firstly, their size does not distort the proportions of other visual elements. Secondly, they can easily be moved around the dashboard.

This is what we did to "free" the cards from the grid to make our work more convenient:

1. Created cells with formulas for calculating indicators in the desired format under the data sample with KPIs.

2. Added "Rectangle" shape to the upper part of the dashboard and added the card title to it.

3. Added "Text box" element to the rectangle and placed a link to the new sample cell in it.

4. Grouped the parts of the card, uniting "Rectangle" and "Text box" into one element.

5. Copied this element and changed the cell link for the title. We also corrected the displaced formatting manually.

6. Changed the default card style by choosing a dark font against light background.

7. Aligned the cards on the dashboard in three steps.

4.3 Designing Interactive Slicers

Removing excess elements is primary to dashboard design. We have already practiced this with charts and KPI cards. The workspace looks much "cleaner," while the content is immediately visible.

The slicers look overloaded against the background. We will get rid of excessive contours and set up correct operation for the interactive report.

How to Disable Slice Borders

We need to disable slide borders to get rid of excessive elements. This is what our dashboard looks like at this stage.

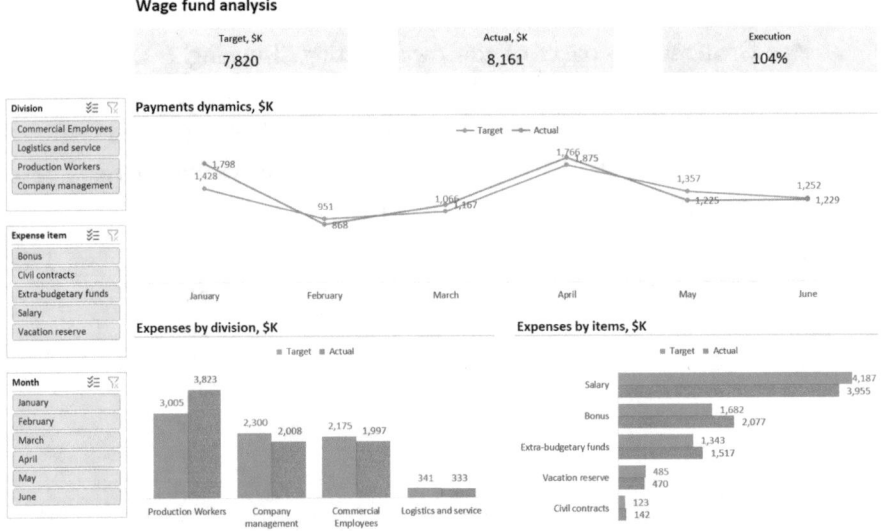

Excel offers a variety of different styles for interactive slicers, but no options without a contour. Let us approach this in a different way.

Step 1. Select any slicer with the mouse and right-click on the light blue style in "Slicer" menu tab ("Options" in older versions of Excel). Select "Duplicate" in the context menu.

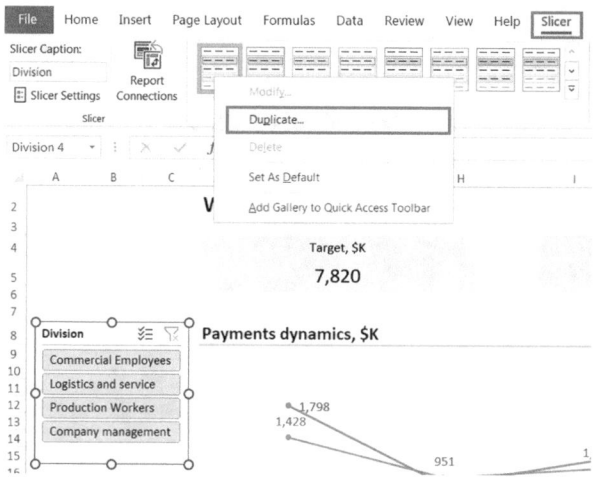

Step 2. Select "Whole Slicer" in "Modify Slicer Style" dialog box and click "Format" button.

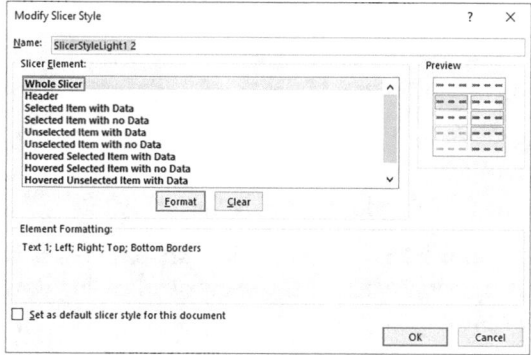

Step 3. A new window "Slicer Element Format" will open to configure the parameters: select "None" on "Border" tab and click "OK."

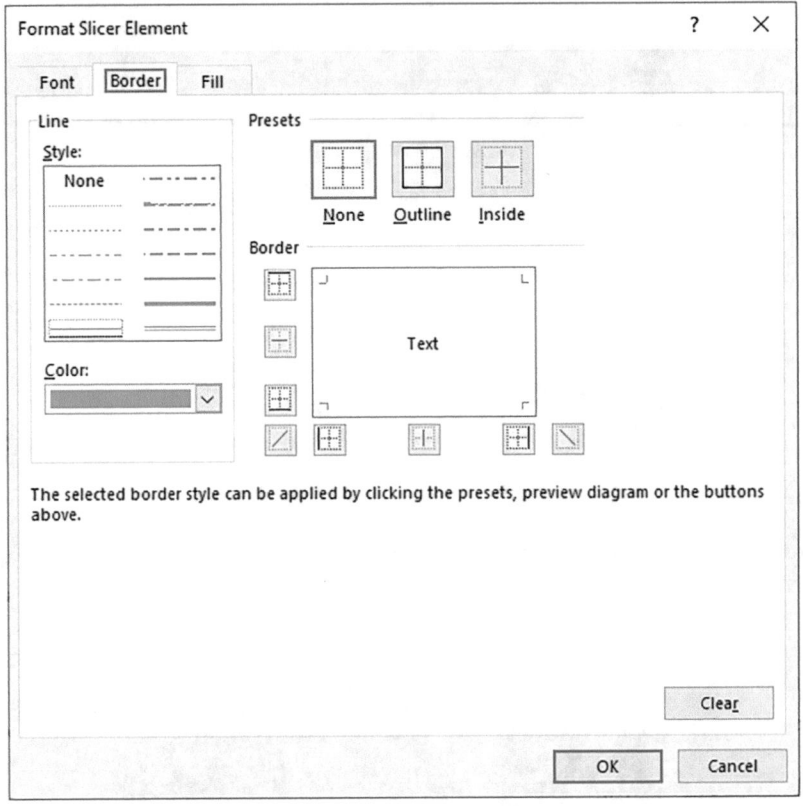

Repeat "OK" in the first window "Modify Slicer Style."

Step 4. A new contour-free option for slicer styles appears in the drop-down menu. To use this option, select all the slicers by holding down Ctrl (or cmd) key and click the option.

Now our dashboard is completely free of excessive lines and visual barriers. We also have extra workspace: you can stretch the slicer over several months and get rid of yet another unnecessary element, the scroll bar.

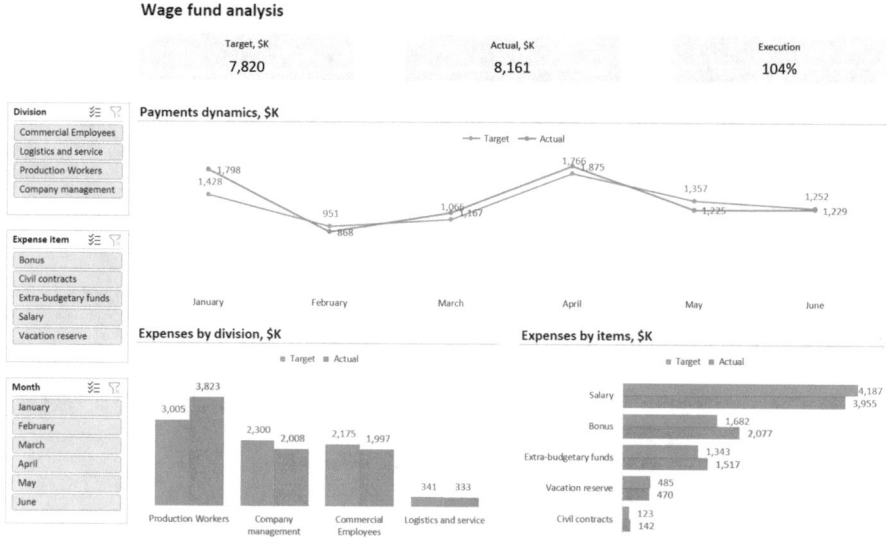

Working with slicers and empty diagrams: avoiding mistakes.

Each interactive slicer on Excel dashboard filters all charts by default – including the chart created on their basis. This looks like an error to the user.

Clicking on month, for example, we only see blurry dots in the section instead of "Expenditure Dynamics" graph.

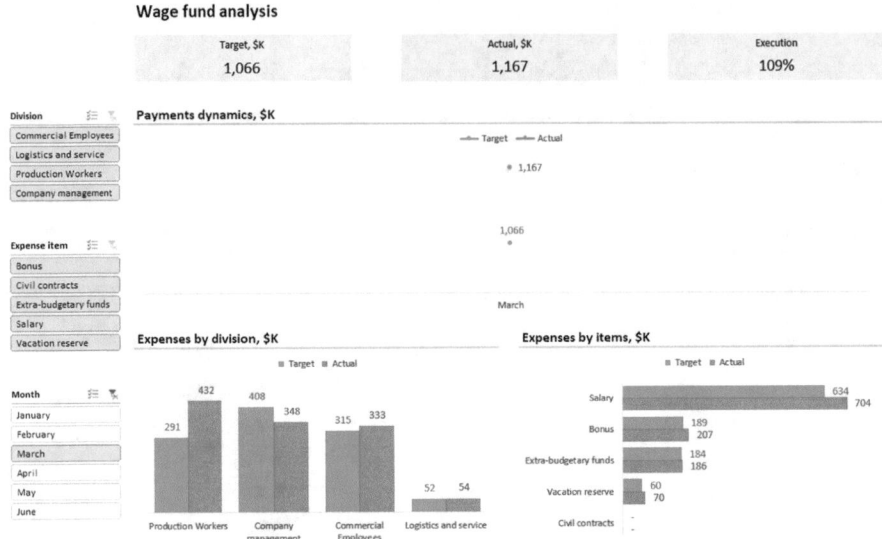

The same happens when choosing a specific division or expense item: the diagram associated with the slicer used does not explain whether it is good or bad relative to the other categories.

If you apply several filters at the same time, you might have a completely "empty" dashboard, which will confuse the user.

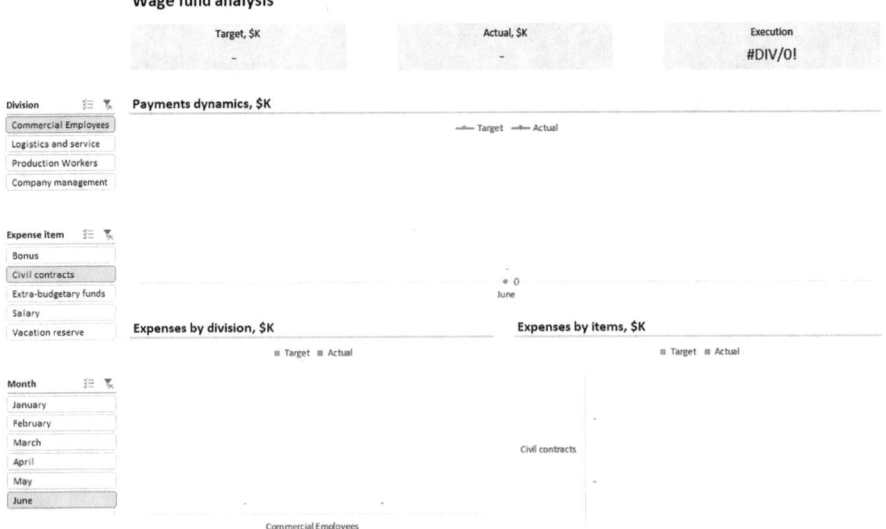

Solving the Problem

To preserve the necessary information when filtering, we will disable slicers on the charts linked to them. I call this "Disconnecting slicers from themselves." We add this setting at the last stages of dashboard development.

We will need a window with a list of reports connected to the slicer. It can be requested in two ways:

1. Select a slicer ➤ open the context menu clicking the right mouse button ➤ select "Report connections."

2. Select a slicer ➤ go to "Slicer" menu tab ("Options" in older versions of Excel") ➤ select "Report connections" button.

We will see the slicer we are setting up in parentheses in the title of the opening window: this is why we gave names to pivot tables in Part 1.3. We now need to uncheck the box next to the data sample bearing the same name.

For example, if we are setting up a slicer by divisions, we need to uncheck the box next to "Divisions" field. "Expenditure by Division" chart will not be filtered when using this slicer.

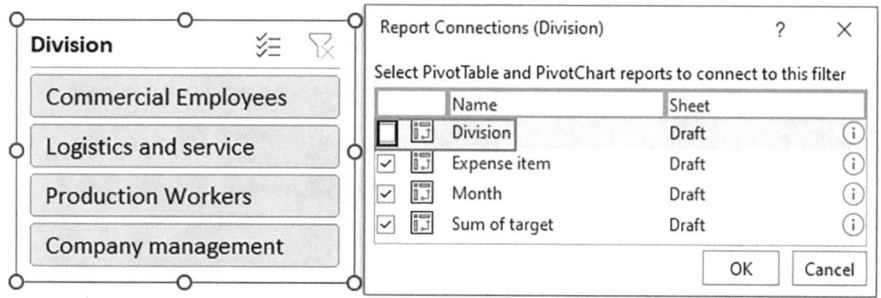

We should perform similar actions with all other slicers: the "by-month" filter should be disabled from "Months" sample, and the filter by expense should be disabled from "Expense Items" sample.

This setting will not allow filters to affect data sets with the same name: you can use any slicer, and all visual elements will remain in their places and will not cause confusion to the user.

Here are some conclusions we can draw applying several filters at the same time:

- The salary of the company management in April took 88% of the planned value of $153,000 – this is what we can see on the cards.

- " Expenditure Dynamics" graph shows how the salary target and actual values for the company management change from month to month.

- We can compare the total salary of the company management in April with the indicators of other divisions in the diagram "Expenditure by Divisions." The salary is lower than that of production and commercial employees, and higher than that of employees in logistics and service departments.

- "Expenditure by Items" chart shows that most of company expenses on management were paid in bonuses: salaries in the division took second place in April.

There are no more empty blocks on the dashboard: the correct slicer configuration makes each diagram as informative as possible.

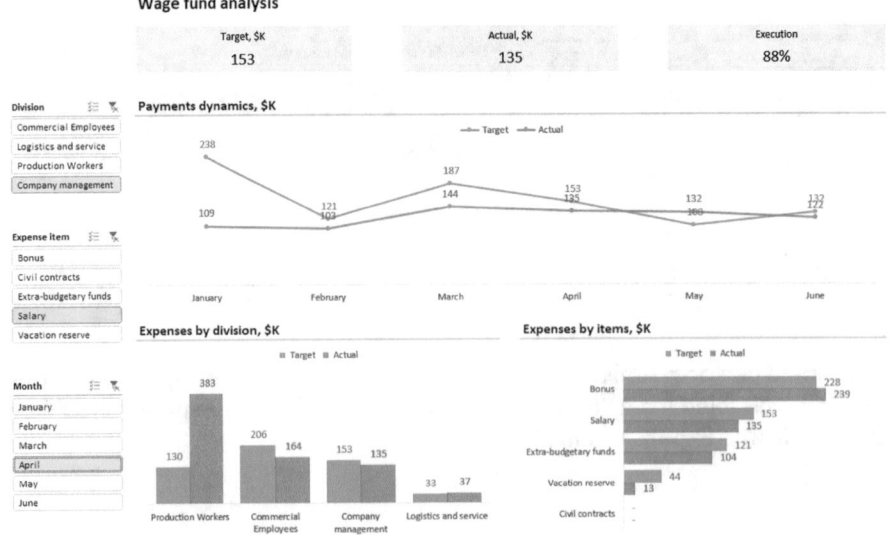

Summary

We customized the appearance and functionality of the interactive slicers at the final stages of dashboard development. This is a quick summary of what we did.

To disable the contour of the slicers, we

- Chose a suitable light blue slicer style
- Changed it by removing the borders in the settings
- Applied a new style to all the slicers

To connect slicers to data samples, we

- Selected "Report Connections" function in the menu
- Disabled the effect of each slicer on the chart linked to it
- Tested the operation of the configured slices on the dashboard

4.4 Working with Colors and Fonts in Excel

When developing a draft version of the dashboard, we used the standard Excel theme. It has its advantages: it does not take much time, and blue and orange colors contrast well with each other.

There are disadvantages as well: this dashboard will not look original, much less expensive. We accepted it as a draft version and went on to advanced design. Our task is to create a dashboard visually different from the standard Excel style.

Let us first study the tools offered by the program. What makes them different and how can we choose the right one?

Tools for setting the "final" parameters

- Palettes – these are sets of colors that combine with each other well (in the words of MS Office creators).

- Fonts – any selection will change the font on all elements of the dashboard.

- Themes – each theme includes a set of "harmonious" colors as well as a set of fonts.

All of them can be found in "Page Layout" menu tab in "Themes" section.

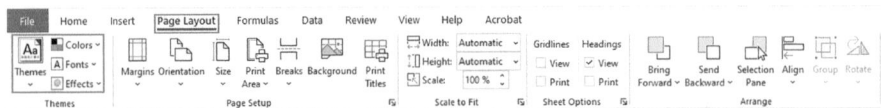

How to Work with Palettes

This option is suitable if you are satisfied with the fonts and font sizes on the dashboard, but we still need to avoid the standard color design.

You can view all the options available by clicking the "Colors" button – there are more than 20 combinations in the drop-down list. You can see how each of them looks on your dashboard by hovering over the palette.

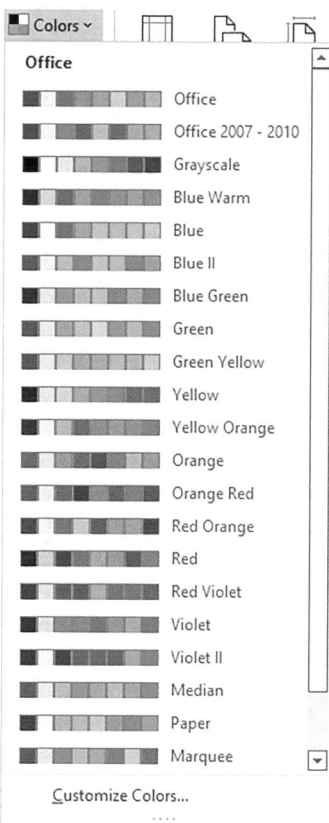

Unsuccessful Similar Palettes

Among these options are many monochrome, non-contrasting combinations. They do not focus on the data and turn the dashboard into a monochrome canvas, and I would not recommend using them. Here are several palettes of this kind.

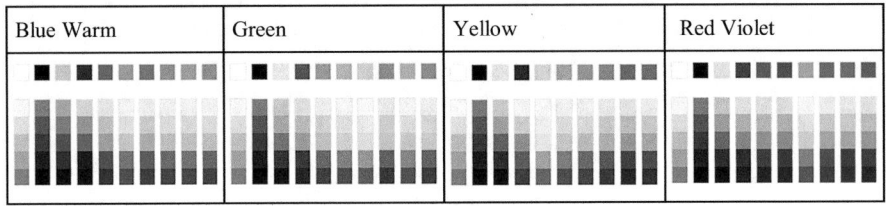

Another disadvantage is that there is no set of colors for visualizing devia-tions. We use red to focus on negative indicators and green to focus on posi-tive ones, while yellow helps to mark values close to critical. Non-contrasting Excel palettes have one or two of these colors.

This is what a dashboard will look like if you use "Red and Purple" palette.

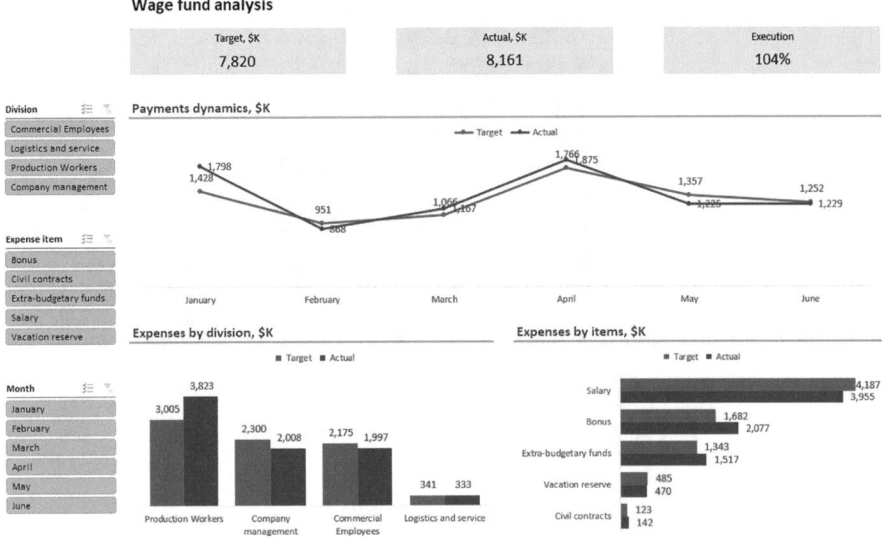

Nothing catches the eye in this color combination: there are no visual accents. Even "Target" and "Actual" are not different from each other, although there should be a clear contrast made between them. The purpose of any diagram is to simplify the comparison making it visual.

Suitable Palettes

There are good palettes in Excel, too. It is quite possible to develop a decent design by introducing minor improvements. The following are some examples.

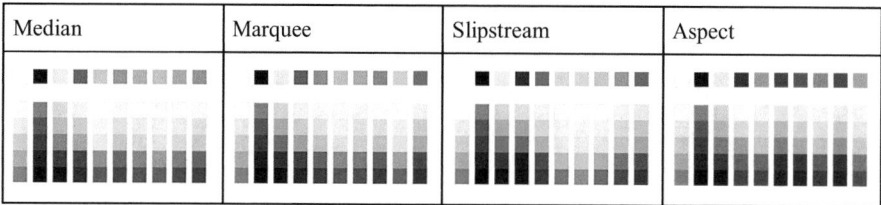

Median	Marquee	Slipstream	Aspect

They are already more balanced, and there is a set of colors to focus on deviations with several combinations for the desired contrast.

You do not have to choose the option in which Excel can color all the elements when choosing a palette. They can all be modified: you can change the color of the cards, slicers, columns, and lines of the graph. Select the desired element and use "Fill color" button in the main tab of the ribbon or in the context menu.

To draw an example, I will focus on "Aspect" palette. It has several disadvantages as listed in the following:

- The background card color, although not bright, still conflicts with the text.

- There is not enough contrast between the indicators of the target and the actual, and the data labels for these categories are barely visible.

- "Actual" columns are red, which is perceived as a deviation and negative indicator.

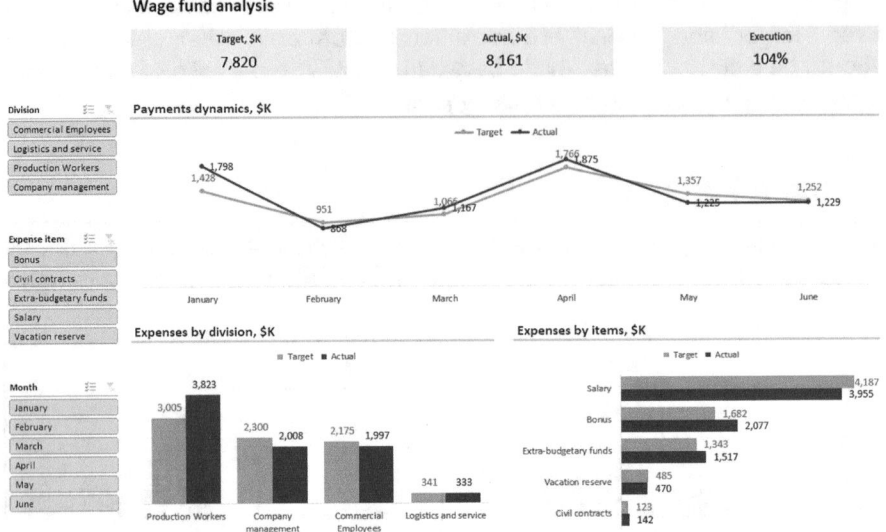

Let us try and edit the original version, working around its faults. This is what I did.

1. Selected contrasting colors for "Target" and "Actual" parameters, so that they are not associated with deviation colors. I chose blue and brown.

2. Chose the lightest shade of the palette - light brown – for cards and slicers.

3. Changed the headlines color from black to dark brown so that they do not attract attention.

4. Selected a harmonious shade of brown for heading dividers.

5. Reset the data labels a shade darker than that of the columns and lines.

The result is a dashboard in a calmer color scheme with an emphasis on data made through the contrast of "Target" and "Actual."

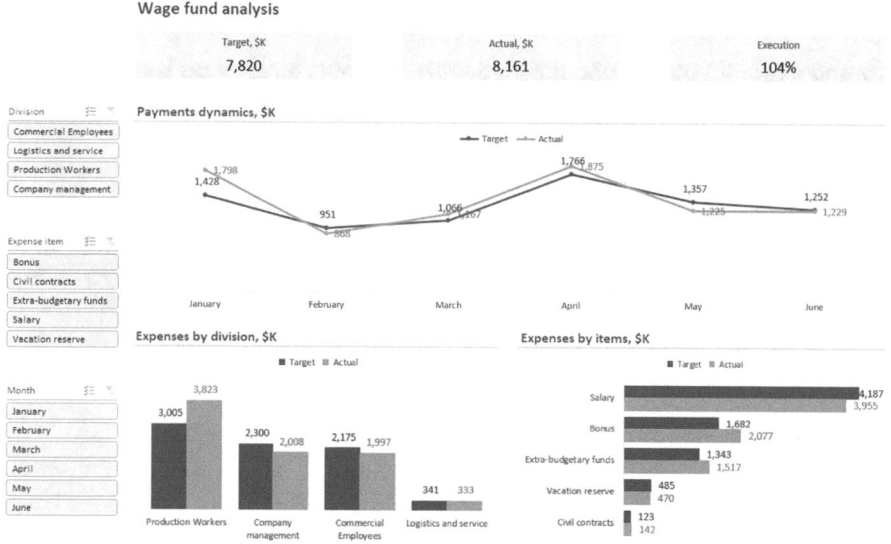

Working with Fonts: Important Points

If you want to "play with fonts" like a real designer, you can also do this in "Page Layout" tab in "Themes" section: there are more than 25 options in the drop-down list.

It may seem that fonts present no difficulty. However, they affect both the appearance and the overall perception of the dashboard.

Please note: the font changes automatically on all elements of the dashboard, except the labels and indicators in KPI cards. Microsoft failed to get even these small things right. You will have to change the font in the cards manually. These small things distinguish professional performance from amateur work.

Fonts for inscriptions and titles can be different. It is important that the caption should have the "cleanest" font, without serifs and other decorations.

Another important rule is to use the same font size for data labels and inscriptions on all charts and graphs. If we chose 12 for a graph with expenditure dynamics, it should remain the same for all visual elements.

■ *Tip!*   *Avoid fonts with bold lettering, serifs, high letter density, or small letter spacing. The font should not attract too much attention. It is better to choose traditional Arial, Calibri, or Tahoma.*

By and large, if you are not satisfied with the fonts already on the dashboard, I recommend going straight to the theme selection in the same menu.

Working with Themes

These are design options that have preset colors and fonts. Any theme can be modified and worked on. Let me demonstrate how we can choose a less time-consuming option.

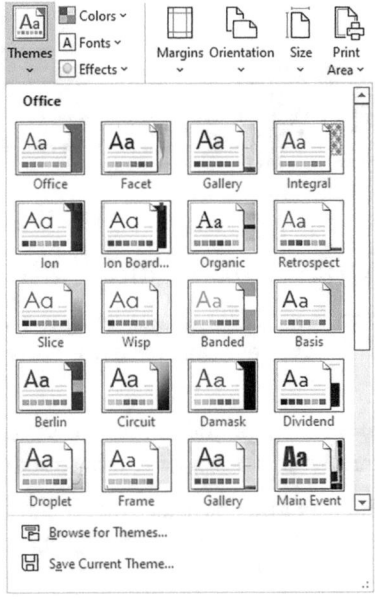

When choosing a theme, pay attention to the font of the inscriptions: in some Excel options, this parameter can negate all the advantages of your dashboard. Here is what will happen if you apply "Main Event" theme.

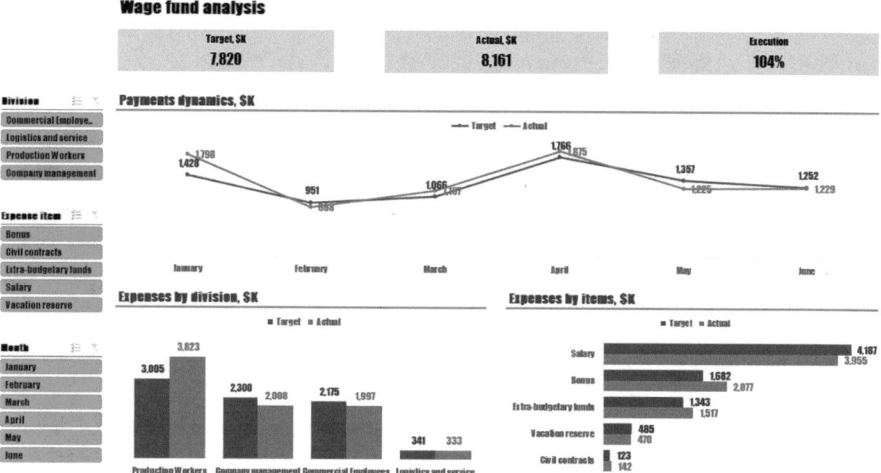

I will be blunt: everything is just awful. An. Absolutely. Useless. Pattern. Senseless and merciless. I hope someone at Microsoft starts hiccupping every time I see these options. Both the colors and the font are useless. I will try to pull myself together and analyze this miracle of Excel design with as little emotion as possible.

Technical Fault Analysis

- The bold font of the card headings distracts attention from the data.

- Category labels on the histogram do not fit horizontally – they are lost due to the same bold font.

- For the same reason, it is the legend that strikes the eye.

- Black color on pink cards and – even more so – on dark red slicers – makes them hard to read.

- Red color is associated with negative indicators: it feels like the company is "in the red."

To refine the theme, you can change individual colors or use a different palette. You can also change the font for all the elements on the dashboard. We will not do this with the theme "Main Event": it will take too much time to correct its faults.

■ *Tip!*   *When choosing a theme, give preference to contrasting options: shades of the same color make information difficult to read and never give emphasis to the data needed.*

Changing colors of the elements in the theme. Select an item and choose a new color in the context menu or in the main tab of the ribbon using "Fill color" button.

Changing color combination in the theme. Select a different palette in "Page Layout" tab in "Colors" drop-down list. The new color combination will be automatically applied to all the elements on the dashboard. The font of the theme will remain the same.

Changing the theme font. In the same menu in "Page Layout" tab, select the option required from the drop-down list under "Fonts" button.

Please note that some themes have different fonts "embedded" for text labels and numbers. Their outlines are usually similar, but the thickness and density of the characters may vary so that the numbers remain visible and do not get lost.

These options are included in "Badge" and "Parcel" theme.

Text font - Corbal	Numbers font - Gill Sans MT
ABCDEFGHI JKLMNOPQR STUVWXYZ abcdefghijklm nopqrstuvwxyz 0123456789	ABCDEFGHI JKLMNOPQR STUVWXYZ abcdefghijklm nopqrstuvwxyz 0123456789

Adapting Template Diagram to New Theme

By default settings, the color theme selected (as well as the palette) does not affect the diagram created from a template. "Expenditure by Items" chart can serve as an example.

To repaint it properly, we will repeat the steps for working with the template from Part 3.5.

1. Save the customized "Expenditure by Departments" chart as a template without changing the path to "Save" directory.

2. Select "Expenditure by Items" visualization, choose "Change chart type" in the context menu and specify the option required in the Templates folder.

3. The template will turn a diagram into a histogram. To fix this fault, repeat the "Change chart type" procedure and choose the chart required from the menu.

4. Adjust the sorting on the bar chart so that "Target" and "Actual" stand in the correct order both in the columns and in the legend.

Summary

In this part, we temporarily moved away from working with the dashboard to get acquainted with the tools Excel offers for the "final" design.

- To select colors and fonts for the dashboard, use "Themes" section in the "Page Layout" menu tab.

- If you are satisfied with the fonts on the dashboard, but still want to move away from the standard color design, test different palettes in "Colors" button.

- Make sure that the palette selected has red, green, and yellow colors to accentuate deviations.

- If you need to change not only the color combination but the font as well, use the themes in "Font" button in the same menu section.

- Make sure that there are contrasting colors in the theme selected: they will help draw attention to the data.

- If necessary, change palettes and theme colors for individual elements or the entire report.

4.5 Improving Standard Excel Themes

In matters of design, there is no correct opinion, and it is often the client who has the final say. That is why my team always offers the customer several options of dashboard design: a minimalist version against a white background, contrasting dark, and a middle color.

In this part, I will show how to create quick design options in Excel based on standard themes. We will work with the styles that customers choose most often – minimalist light colors and the middle color with bright accents.

Example 1: Primary Colors Should Be Contrasting

Let us take theme "Facet" as an example. We have already considered the palette by the same name, but the theme looks completely different. The default design here is different shades of green.

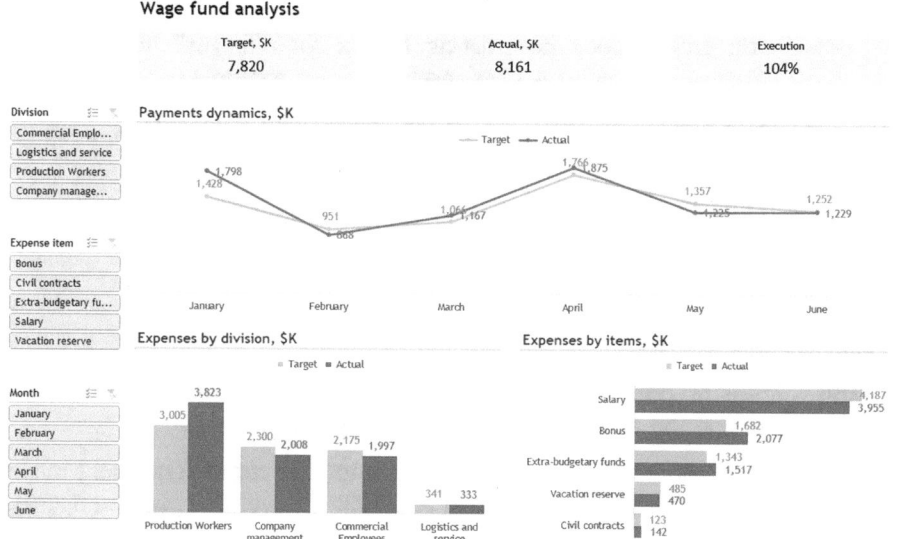

Why the Theme Is Suitable

The default result may not seem appropriate. However, if we look at the set of colors available and analyze the fonts, we will see that the theme can be refined.

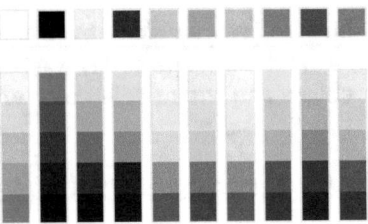

- There are colors to accentuate deviations. Green, red and yellow have good saturation and will be visible against the background of the other colors.

- You can choose contrasting colors for "Target" and "Actual" indicators – blue and brown, for example.

- The palette available allows you to choose appropriate shades for the remaining dashboard elements.

- The font of the theme is sans-serif, the lettering is normal. All the numbers are of the same size – they are easy to read.

Analyzing Faults

Firstly, there is no contrast in this combination of green colors: the columns of the target and the actual merge into one blurry shape.

Secondly, the green color is associated with a positive result. The client might therefore miss the net excess of actual expenditure over the planned figures.

Thirdly, the color brightness in the cards and slicers is close to the fill brightness of "Target" columns. We need to dim the background of service elements, which will accentuate data visualization.

Fixing the Problem

We will select light shades of green and yellow for cards and slicers so that they do not attract attention.

We will assign a dim dark gray color to the titles so that they fit better into the new range.

We will select a contrasting shade for "Target" columns and lines so that they differ from other elements. Here the shade is yellow.

We will leave the default color green for "Actual" columns and lines.

We will set data labels to colors a shade darker than those of the corresponding columns to make them more visible.

The result is a completely different dashboard made from the same color theme "Facet."

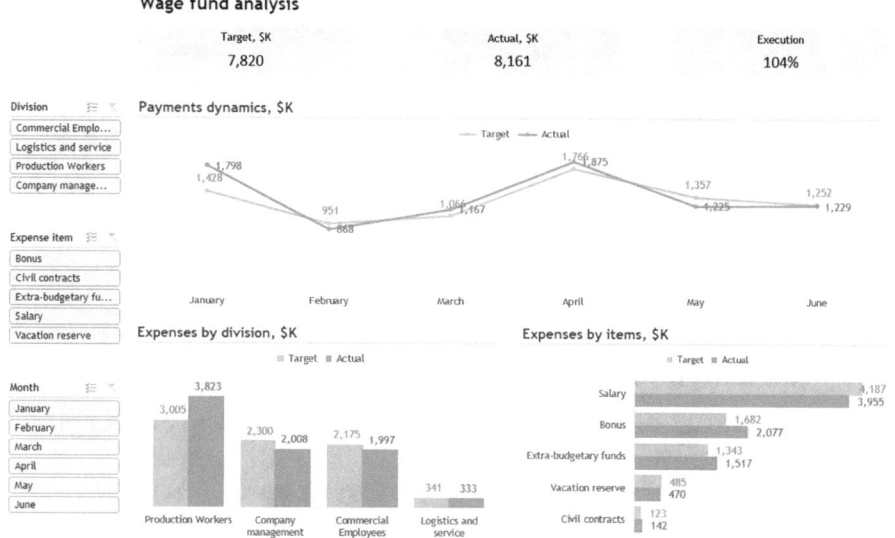

Example 2: Less Brightness for Service Elements

Let us consider another option from Excel – the "Badge" theme. It provides a good palette, but it has default excess of yellow, and the saturated color of the slicers distracts attention from the data.

Why the Theme Is Suitable

The palette shows the choice of colors we need to complete the design.

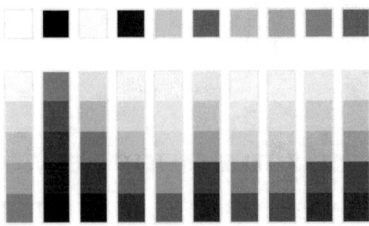

- There are colors to accentuate positive and negative indicators: green, red, and yellow.

- There are two options for the primary colors. The first option is "golden" for "Target" and olive for "Actual." The second one is blue green for "Target" and purple for "Actual." Each combination will have the right contrast.

- The theme has a good font – there are no serifs or decorations on the letters, the numbers are even and easy to read.

Analyzing Faults

You can find a few minor flaws in this theme, but I suggest we focus on its greatest fault. A large amount of yellow on all the elements distracts attention from the data – you need to try hard to locate necessary information on the dashboard.

Fixing the Problem

- Change the color of the cards and slicers to remove the yellow saturation. I made them less bright, setting a universal light gray shade.

- Choose a different color for dividers under the chart titles from the theme palette. I chose light purple.

If we change the colors for the indicators to blue-green and purple and invert the colors on the cards, we will have a completely different dashboard.

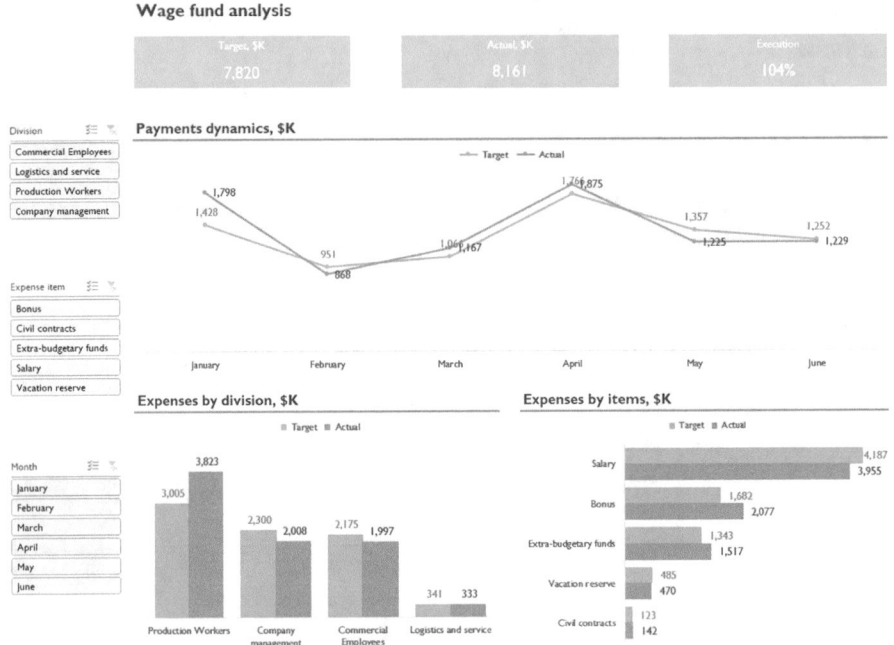

I mentioned that I recommend using a light background and bright dark font on the cards. The option with a rich fill, however, looks more original and some clients might like it better.

Demand creates supply. Sometimes clients ask for "gold on black with a decorative pattern," because they have never witnessed other examples of "expensive" design. This is also why we offer a choice of several design options.

Working with the dashboard daily, the client understands that a black font against a light background is more convenient to read.

Example 3: The Background of the Dashboard Does Not Have to Be White

Let us look at another option from Excel, "Parcel" theme. There is a lot of yellowness here too, but even in its default version "Target" and "Actual" contrast well. You can leave it as it is, or you can spend another half hour and exceed the customer's expectations.

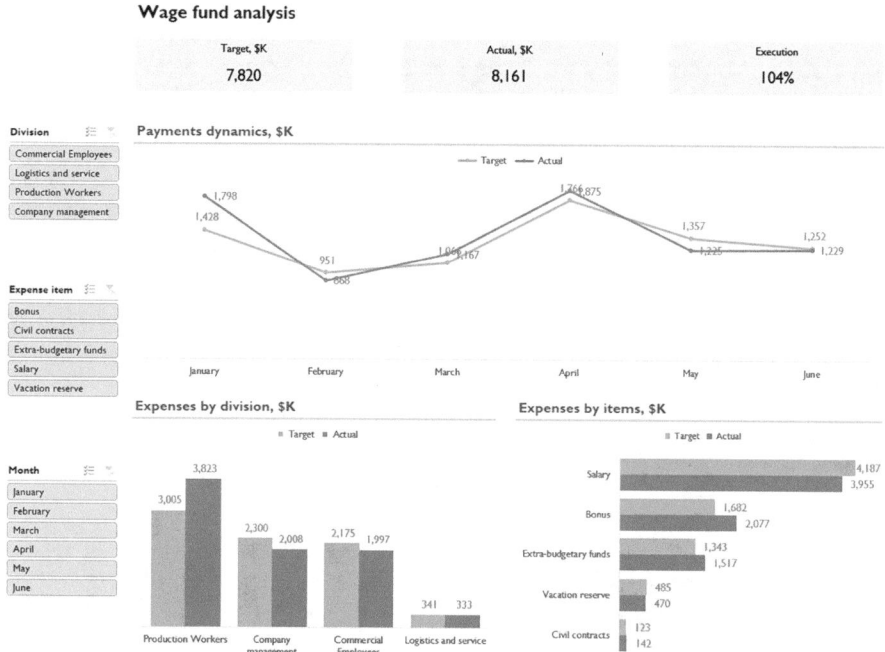

Why the Theme Is Suitable

At first glance, the palette may seem too pale due to its neutral pastel shades. You can still create an original and functional dashboard style using these shades. The palette offers interesting shades – brick, olive, and gold – in place of the classic "traffic light" (green, yellow, and red).

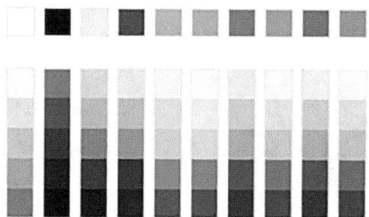

Analyzing Faults

The theme itself has no obvious drawbacks, but working with it, you can arrive at colors that are too pale or that merge with each other. If you do not trust your intuition to choose competent color combinations, do not take risks.

Here, for example, the theme was an unfortunate choice for the dashboard, and everything turned yellow again. This is what we will fix.

Fixing the Problem

We decided to move away from the standard Excel layout – now we need to remember that a white background is only but an option. You can use any color as long as it is not too bright.

To change the background color, I select all the cells by clicking on the triangle in the upper left corner of the Excel sheet, just above the row numbers. I choose the lightest shade of gray for the fill. The fill of the chart area will remain white.

I choose light blue instead of yellow for the slicers, and this immediately makes our dashboard calmer. I assign dark blue to the card background to make the cards more visible.

I will set the contrast between the columns and lines "Target" and "Actual" in other colors – brick and olive, respectively.

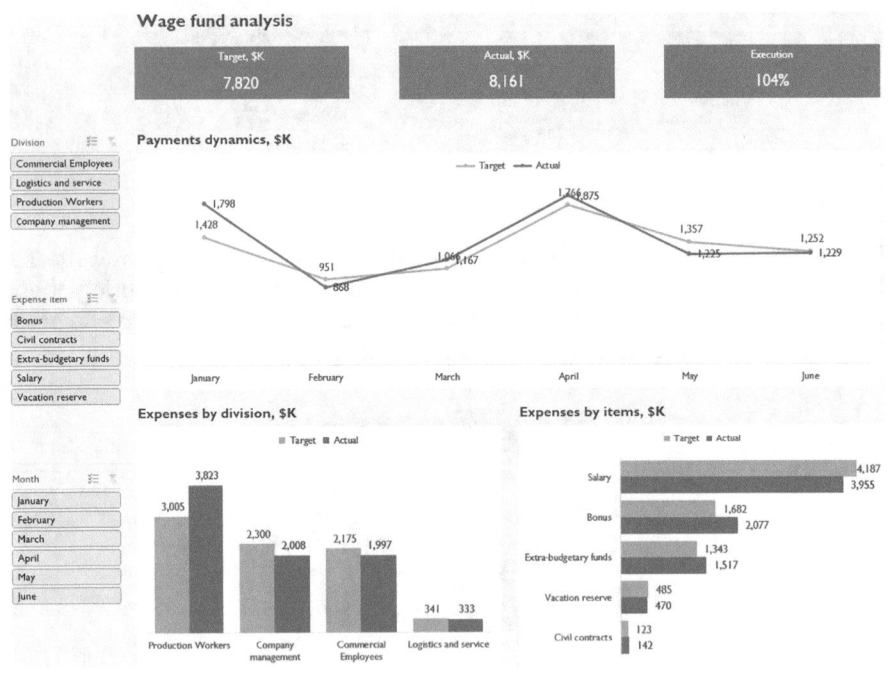

Note I removed the lines under the titles, which created clear boundaries between semantic blocks. Now this function is performed by a white background of diagrams, which means that additional lines are no longer needed, it is not necessary to overload the interface with them.

Summary

We figured out how to make a quick dashboard design in different styles and offer a choice to the customer. We also received several new options from the usual design in blue and orange colors.

- .We changed the colors in "Facet" theme and changed the merging green dashboard into bright green and yellow.

- We created two variants with "Badge" theme in different colors: calm gray with an emphasis on data and blue green with purple.

- We changed the workspace color in the "Parcel" theme and chose rich olive and muted brick for the main indicators.

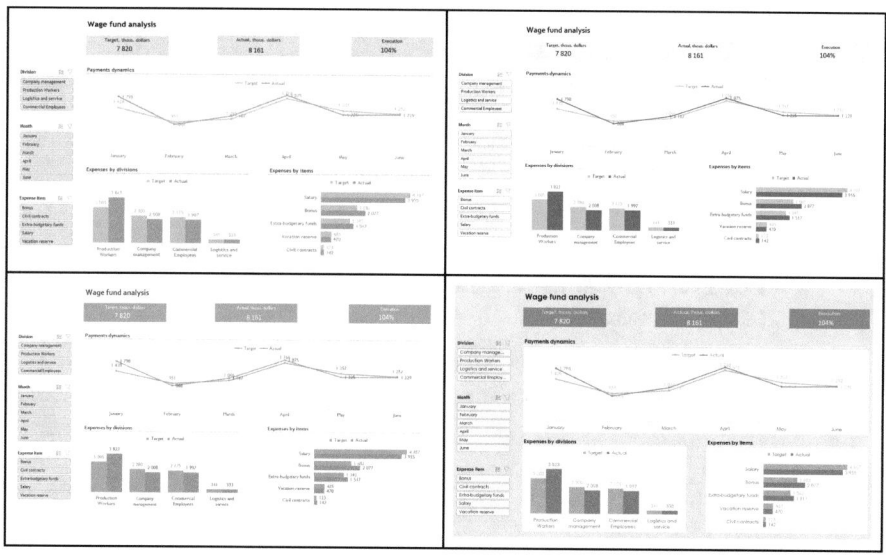

Experiment with improving the standard themes. Do not hesitate to show the results to the customer: ask them to choose something and explain their choice. This will help make the upcoming dashboard versions more suitable for the user.

Color Theme Adaptation Checklist

I suggest the following checklist to select colors on your own. The rules apply both to Excel and other data visualization tools.

1. The main colors of the palette are contrasting. There are shades of green, red, and yellow. Or at least two of these colors.

2. The background of the KPI cards does not conflict with the text. Choose a black or dark gray font against a pale background. Sometimes a white font and rich background – blue, for instance – will look good too.

3. Chart columns do not merge with each other. Choose contrasting pairs, for example, blue and orange. Or make one accent: bright "Actual" and gray "Target."

4. The data labels are clearly visible. Their color should be a shade lighter or darker than the column or line of the graph – do not let the values get lost.

5. The background of service elements, for example, slicers, is dimmed. Gray is universal. Or the lightest shade of the main theme color.

6. Additional graphic elements (backgrounds for diagrams, titles, dividing lines) are paler – they should not attract too much attention.

7. The main text in the diagrams (data labels, axes) should be uniform. The only exceptions are titles and cards.

Corporate Identity

We figured out how to offer the customer design options that differ from the standard one. But large companies already have an approved corporate style, and the dashboard must match it.

It is not difficult to change the fonts and color of the diagrams – this will already bring 80% of the desired result. But the professional and the amateur are just distinguished by the remaining 20% of the details.

In this part of the book, you will learn the fine-tuning of dashboard elements. Thanks to them, some customers do not even realize that this is an ordinary Excel, and not an expensive BI system. The business customer will appreciate such work much higher.

© Alex Kolokolov 2023
A. Kolokolov, *Make Your Data Speak*,
https://doi.org/10.1007/978-1-4842-8942-6_5

5.1 Creating a Theme in Accordance with the Brand Book

Many companies have templates of corporate-style reports. Most often these are traditional presentations that already contain the right set of colors, fonts, and other elements, for example, a logo. But sometimes there is a corporate style, but there are no documents from which it can be taken yet.

In this part, we will work with both options. First, we will learn how to create our own theme in Excel for the dashboard design. And then we use a ready-made template of the theme from the presentation in PowerPoint.

How to Create a Theme from Scratch

Of course, you can manually change the color of each element on the dashboard and set the desired font in each of them. But it will take a long time. I will show you how to save this time and create a "corporate" theme in Excel as quickly as possible.

I will demonstrate the process using the example of the brand book of the Institute of Business Intelligence. Here are the elements of corporate style from which we will form a new theme:

- Color for light background: light yellow #FFFCEB
- Color for dark background: dark brown #1F1A1A
- Brand Colors: yellow #F8C845, brick #CB5033, blue #2C4A94, green #008864
- Font: Montserrat

Let's set up these colors in Excel, add the desired font and save the entire set as a new theme.

Step 1. On the "Page Layout" tab in the upper-left corner, select the "Colors" drop-down menu and click the "Customize Colors…" item in it. The window "Creating New Theme Colors" will open – in it we will build our palette.

Step 2. By clicking the triangle in each line, we open the menu and select "More Colors…" in it. A new window opens in which we set the code of the desired color. Thus, we specify the colors for the dark and light background, and set the rest in the "Accent" lines.

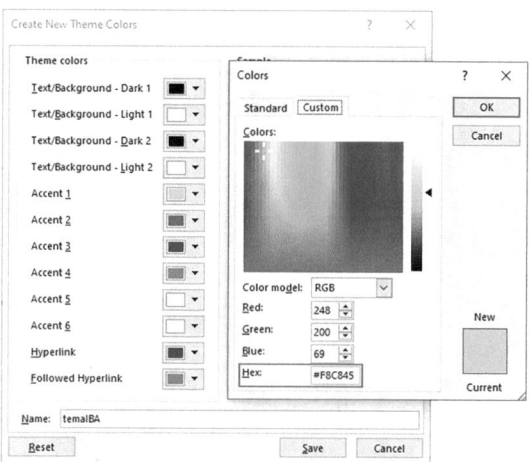

Step 3. When all the colors are configured, in the "Create New Theme Colors" window, we give a name to our theme and click "Save."

Step 4. Go to the font settings. On the menu tab "Page Layout," select "Fonts" ➤ "Customize Fonts…." Enter the name of the Montserrat font, specify the name of our new theme in the following line and click "Save."

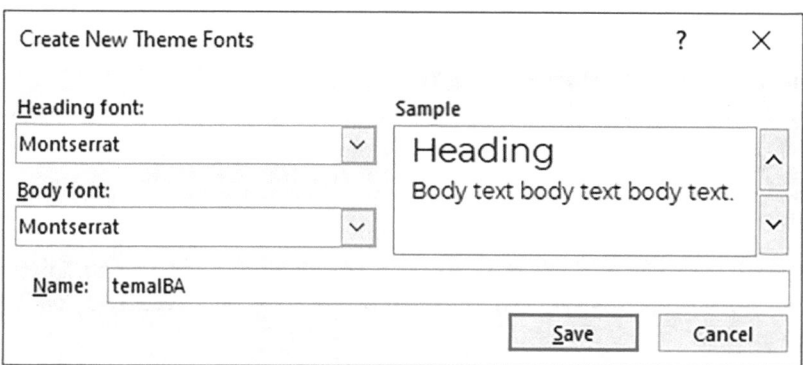

Step 5. When the colors and fonts are configured, we save the new theme. To do this, on the "Page Layout" tab, click "Themes" and select "Save Current Theme…."

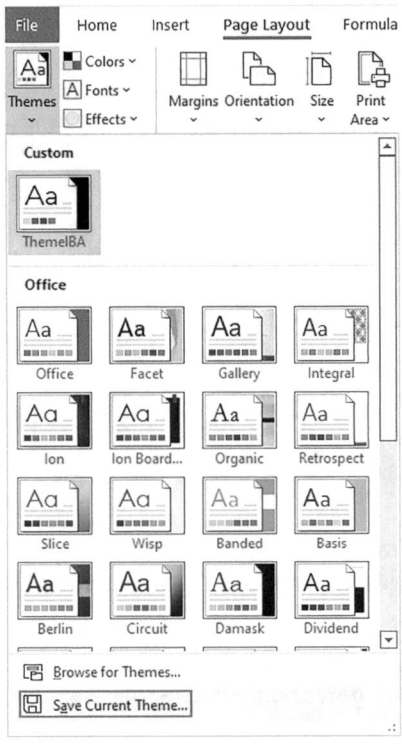

Done! A new section with custom themes has appeared in the list of available themes, where our "branded" one is located.

Life Hack: Importing a Ready-made Theme from PowerPoint

This part is useful for cases when the company already has templates of corporate-style presentations. They can be used for an interactive dashboard in Excel.

I will show you how to do this using the example of the presentation of the Institute of Business Intelligence: it already has the right set of colors and approved fonts of the company. According to the same scheme, you can use the dashboard design of your organization.

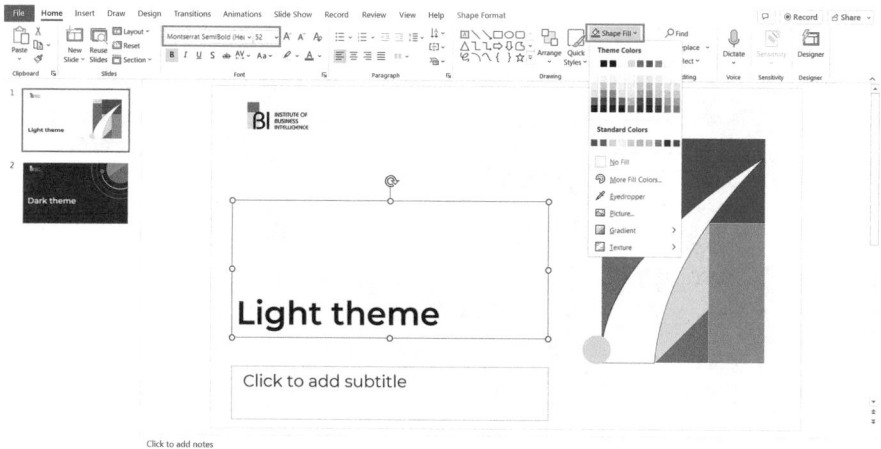

A corporate theme from PowerPoint, by analogy with a diagram template, can be saved on your computer and then used in Excel.

Step 1. Open the presentation, open the "Design" menu tab, open the drop-down list in the "Themes" section and select the "Save current theme" option.

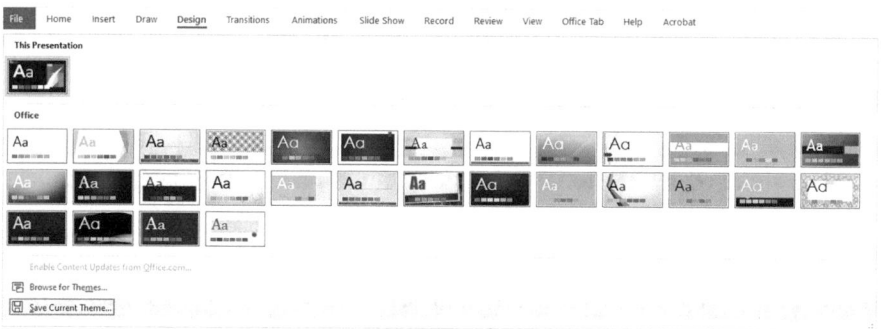

In the dialog box that opens, specify the name of the file, select the folder to save the theme and click "Save."

Step 2. In an Excel workbook, on a sheet with a new version of our dashboard, go to the "Page Layout" menu tab, click "Themes" and after the list with icons of topics, we find the "Browse for Theme…" button.

In the window that opens, "Choose Theme or Themed Document," go to the folder with the desired file, select it and click "Open."

Done! The theme from PowerPoint was automatically applied to the entire Excel workbook. You can check this by looking at the available colors: they must match the palette in the presentation.

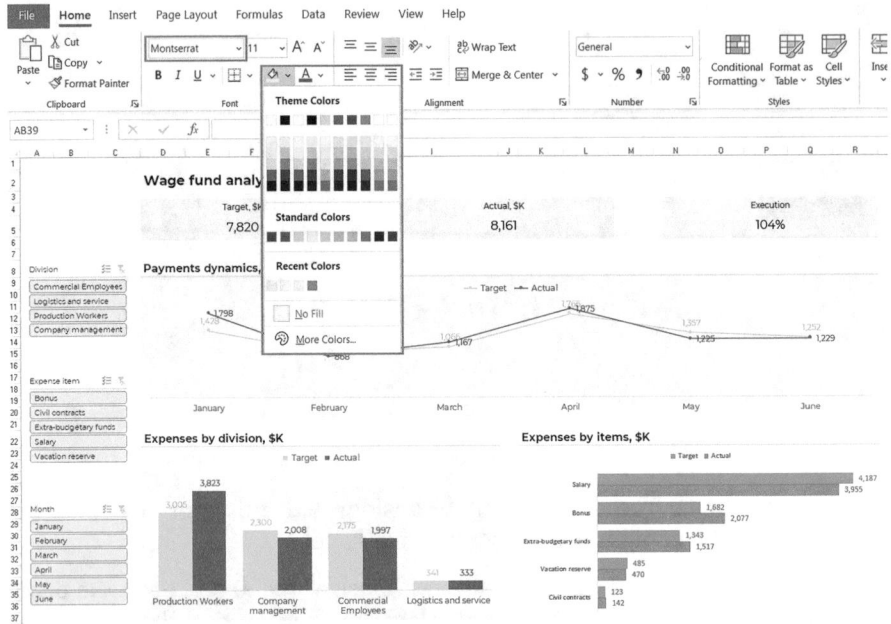

The Montserrat corporate font is wider than the standard one, so the inscriptions on the "Expenses by divisions" diagram are located at an angle. I just increased the size of the chart so that they become horizontal again. If this didn't help you, you will have to reduce the font of the inscriptions to 11.

Here again we encounter a flaw in Excel: the font has changed on all elements of the dashboard, except for the KPI cards. It will have to be changed manually.

As with the built-in color themes in Excel, the new style is not automatically applied to a chart created from a template.

It annoys me a little when one visual element gets out of the general row, so I save and use a new template every time the color changes. Just like we did in Part 3.5, using the "Change chart type" option. After that, we see what the new default theme will look like.

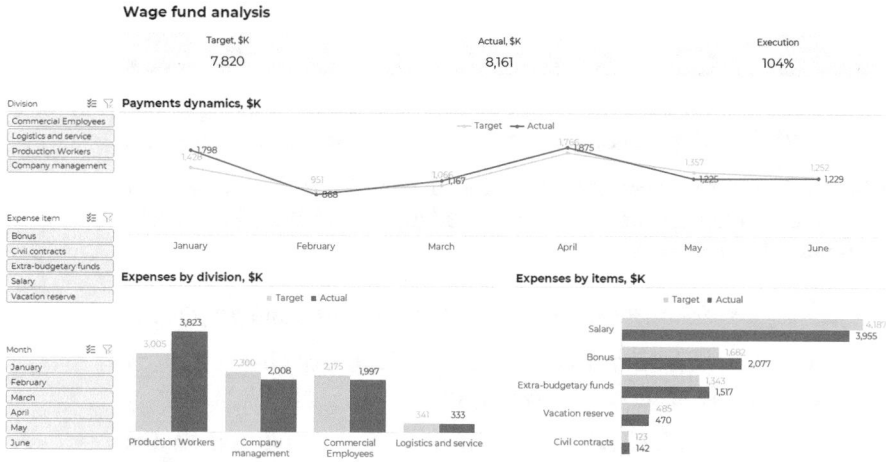

Verification on the Checklist and Points for Improvement

To bring the new theme in line with the corporate identity, let's check the design according to the checklist from the last part. And at the same time, we will add a few more points that we will have to work on.

1	The palette contains a contrasting set of colors, there are shades for the "traffic light"	✓	With this, the theme is in full order
2	The background of the KPI cards does not conflict with the text	✓	Black font on a light background is easy to read
3	Chart columns do not merge with each other	✗	Lost contrast on new backgrounds
4	Data labels are clearly visible	✗	Poorly distinguishable on a new background
5	Background of service elements is muted	✓	Corresponds to the palette of corporate identity
6	The underlays of the diagrams are not conspicuous	✓	Pale color does not distract attention
7	Font and color of data and axes are the same	✓	Now here is the Montserrat font
8	Dashboard background in corporate style	✗	There are white spots between the diagrams and on the slicers
9	KPI cards and indicators on them remained noticeable	✗	Do not stand out from the general background
10	The font of all inscriptions corresponds to the corporate	✗	Not changed automatically for all elements

Summary

There are different ways to design a dashboard in a corporate style. We have reviewed the ones that will help you do this quickly: creating a theme in Excel from scratch and importing a theme from a PowerPoint presentation.

After applying the new theme to the dashboard in Excel, it is important to understand what needs to be improved and corrected. That's what needs to be finalized in our case.

- Restore contrast between chart columns and graph lines.

- Pick new colors for the data labels so that they are clearly visible against the pale yellow background of the charts.

- Tidy up the new background of the dashboard so that there are no white spots on it.

- Make KPI cards and indicators on them more noticeable so that they do not merge with the background.

- Change the font in the cards to that approved in corporate style, because it has not changed automatically.

Download the theme in the corporate style

5.2 Adapting the Theme According to the Checklist

Many complain: you learn to create dashboards, but bosses still require presentations. We have to get out of it, and it doesn't always work out fine.

Someone inserts individual diagrams, and the format falls off on the slide. Someone adds screenshots and moves them for hours so that everything is smooth. And if the data has been updated, everything has to be redone manually…

This part is so that you don't waste time and effort on turning a dashboard into slides. We will make it so that it will replace any presentation, only it will also be interactive.

To get such a result, we need to correct the shortcomings that appeared due to the use of the theme in the corporate style. Some elements "faded" and became difficult to distinguish, others lost the necessary contrast. In this part, we will correct everything that did not fit with our checklist, as well as get rid of new shortcomings and bring the dashboard into full compliance with the corporate identity.

White Spots on the Dashboard Background

The background in the presentation template is not white, but pale yellow, close to the "paper" shade. It was applied to diagrams, but the fill of cells on the sheet remained white. Because of this, the space between the diagrams looks like unpainted stripes.

To fix this, select all the cells of the sheet by clicking the triangle in the upper left corner of the table, above row 1. On the Home menu tab, under the Fill color button, select a pale yellow color (by default, it is the first in the palette).

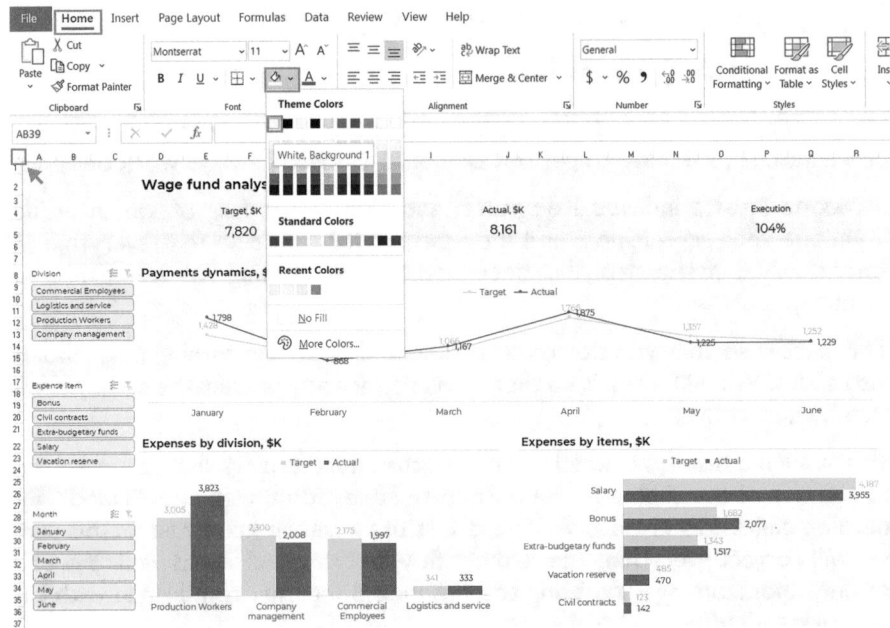

Brightness for KPI Cards

Due to the new color theme, the cards with key indicators were lost in the general background and became almost invisible. To fix this, we will change their background and color, and also check the corporate font again.

- The background of the cards – let's take a contrasting blue.

- Font of inscriptions – Montserrat SemiBold.

- The color of the inscriptions is white.

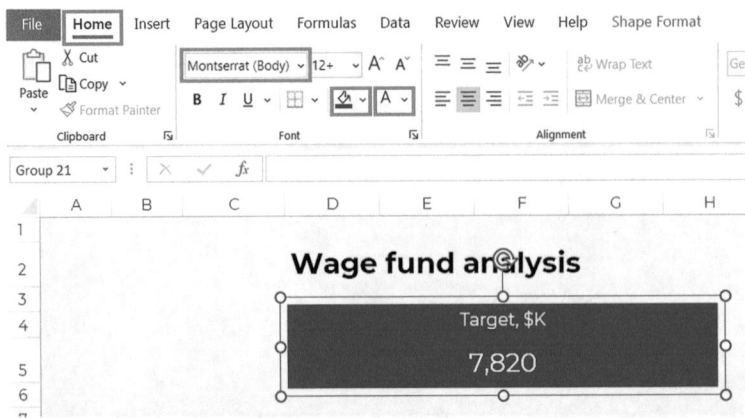

Tip!   *Make sure that the font of all inscriptions has changed to "branded." Excel can show a new font in the font selection field, but when selecting the text, it turns out that the font has remained "standard."*

New Font for All Elements

The font change must be checked on all elements of the dashboard. You should start this check with the title. If the font remains standard, change it to Montserrat and select bold. Also make sure that the font of chart titles and data labels has changed.

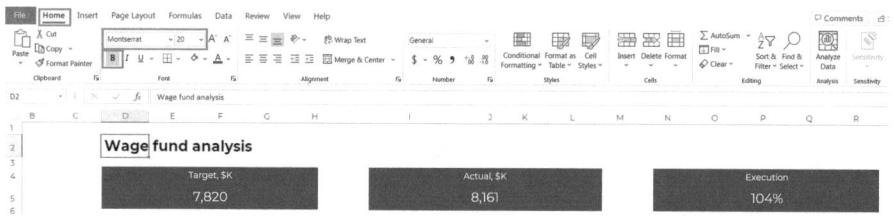

More Contrast on Charts

In the diagrams, the saturated yellow and brick colors are distinguishable, but it's still worth adding contrast. This is especially noticeable on the chart, where a thin yellow line merges with the background. The color for the "Target" category on the graph and diagrams will be replaced with blue, which was used in the cards.

Each chart can be configured manually. But it's faster to do it using a template. After completing the two visualizations, save the bar chart as a template and use it for the item "Expenses by items" – just as we did in Part 3.5.

Remember that Excel sorts the columns of a bar chart in reverse order by default. Change the sorting according to the instructions in Part 3.4.

Don't forget to set a new font color for the data labels. Here we act according to the already familiar rule: choose a shade a shade darker than the color of the corresponding column and line.

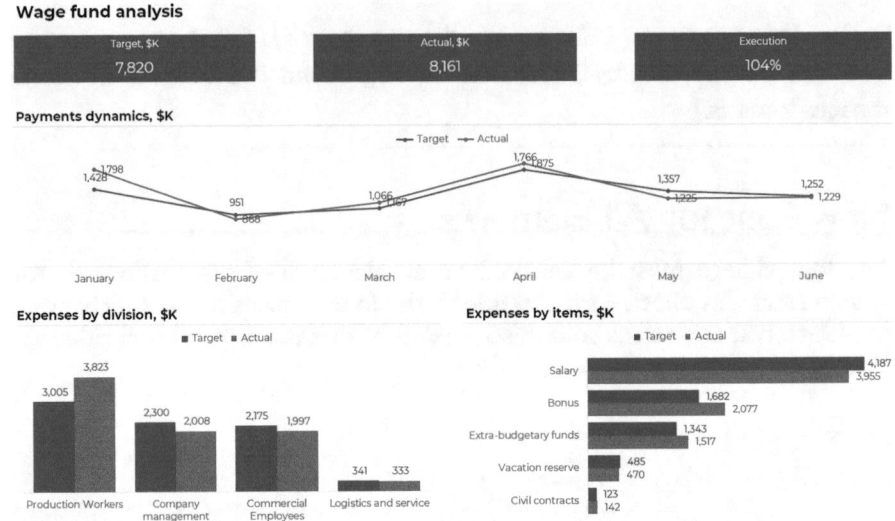

Wage fund analysis

Target, $K	Actual, $K	Execution
7,820	8,161	104%

Payments dynamics, $K

Target —— Actual

1,798 / 1,428 ... 951 / 868 ... 1,066 / 1,067 ... 1,766 / 1,875 ... 1,357 / 1,225 ... 1,252 / 1,229

January　February　March　April　May　June

Expenses by division, $K

Target ■ Actual

3,823 / 3,005 — Production Workers
2,300 / 2,008 — Company management
2,175 / 1,997 — Commercial Employees
341 / 333 — Logistics and service

Expenses by items, $K

Target ■ Actual

Salary 4,187 / 3,955
Bonus 1,682 / 2,077
Extra-budgetary funds 1,343 / 1,517
Vacation reserve 485 / 470
Civil contracts 123 / 142

Adapting Slicers

The background color of the slicers remained white – now they are out of the corporate identity. To change this, select any slicer and go to the "Slicer" menu tab (in older versions of Excel, this is the "Options" tab). In the section "Slicer Styles," right-click the first style in the list and select the "Edit" button.

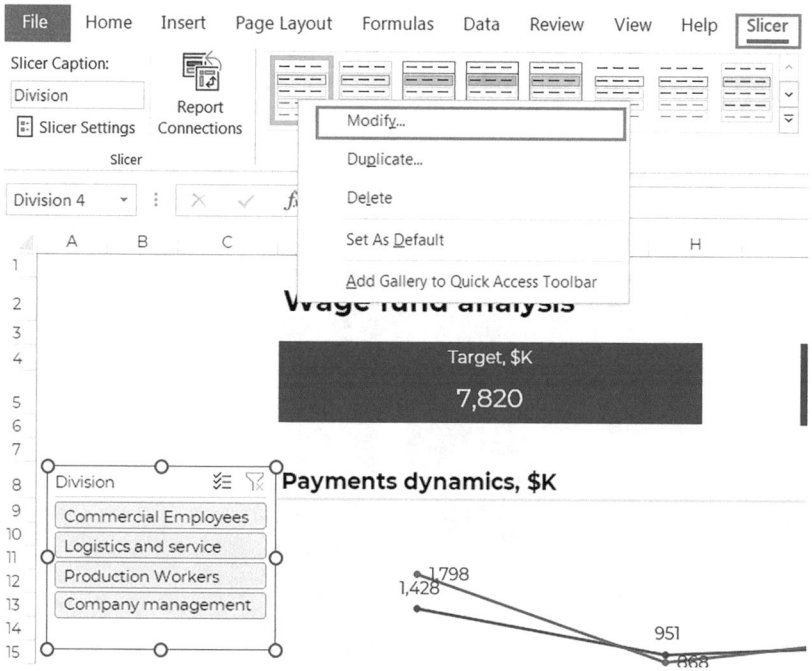

In the window that opens, select the "Whole Slicer" and click the "Format" button. In the new window, on the "Fill" tab, we set a pale yellow background color.

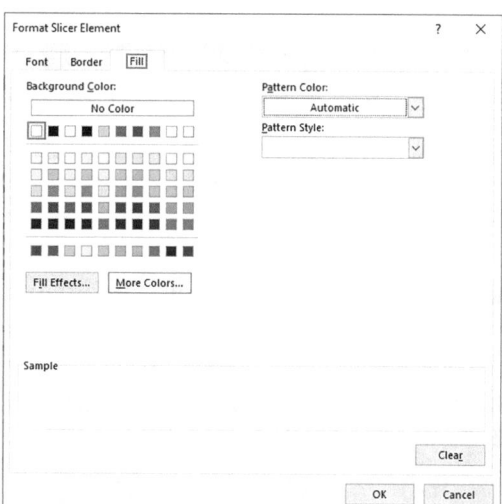

The changed style is automatically applied to the rest of the slicers. We have removed these service elements from the white background and brought them into line with the corporate identity.

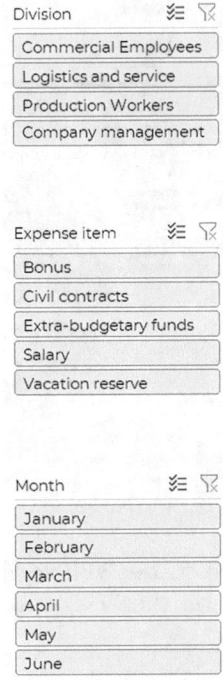

Finishing Touches and Presentation Mode

The last touch remained – the company's logo on the dashboard. There is a special place for it in the upper left corner of the screen, above the slicers. I use the logo from the same PowerPoint presentation, but you can also upload a separate image.

This completes the creation of an interactive dashboard in the corporate style. If you have saved the "standard" version, now you can compare how much the result has changed.

To enhance the effect, it remains to prepare a "Presenter Mode," especially if you will show the dashboard on a large screen.

PowerPoint has a slideshow mode, and many BI systems also have a full-screen demo mode. There is no such function in Excel, but as always, we will assemble it from improvised means:

- On the "View" tab, uncheck the box next to the "Formula Bar" item.

- We do the same in the "Headings" checkbox to remove row numbers and column letters, which means to increase the scale of the dashboard by another 5–10%.

- Double-click the title of any of the menu items or on the "Collapse ribbon" arrow to hide the ribbon.

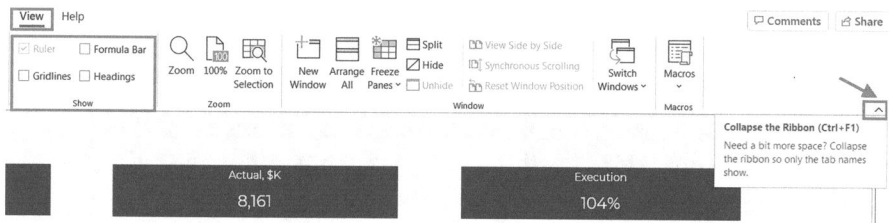

Such a dashboard no longer needs to be inserted with a screenshot into the presentation: now it is itself a beautiful interactive presentation.

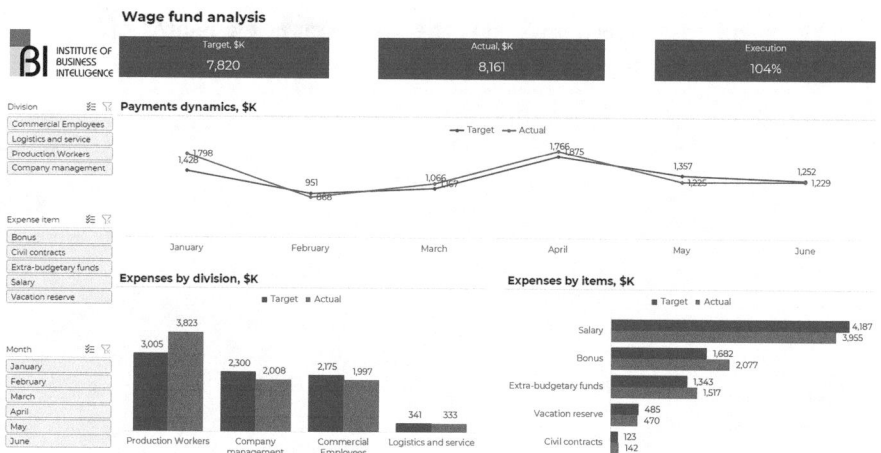

Summary

In order for the dashboard to match the brand book, it is not enough to apply a "corporate" theme to it. New colors can reduce contrast, new fonts can distort the layout.

For a professional result, check your work according to the checklist items from Part 4.5. And do not forget about the elements of corporate identity.

- The background color is correctly distributed throughout the panel: there are no white spots between the diagrams.

- The standard font of titles and inscriptions (including in cards) has changed to "corporate."

- The size of the new font is optimal for the dashboard: the inscriptions are arranged horizontally and are easy to read.

- There is enough contrast between the elements of the diagrams that need to be compared, and they do not merge.

- KPI cards were not lost on the general background because of the theme used: they are clearly visible.

- The background of the interactive slicers is also decorated in the right colors: there are no white spots left on it.

- The dashboard has all the necessary elements of corporate style, such as a logo.

5.3 Creating a Dashboard in a Dark Theme

The dark style looks more original and more expensive both in presentations and on dashboards. But to make it so, you need to make more effort: go through all the elements, make sure that they are clearly visible and contrast enough. Somewhere this contrast will have to be added, and somewhere removed.

Some nuances cannot be put on the checklist – they will require experience and observation. That is why I recommend using a light theme, it is objectively easier with it.

If you have to fit your work into a specific corporate identity, then this part is for you. If not, you can finish the practice and move on to the visualization rules.

I recommend creating a dashboard in a dark theme from a copy of a light one. Copy the sheet with the "branded" version and give it a clear name. This way you will keep both a light and dark dashboard in a corporate style.

The theme that we took in PowerPoint already contains colors in the palette for both light and dark options. Therefore, we use a new fill – the darkest shade of brown. And let's check what turned out well and what needs to be fixed.

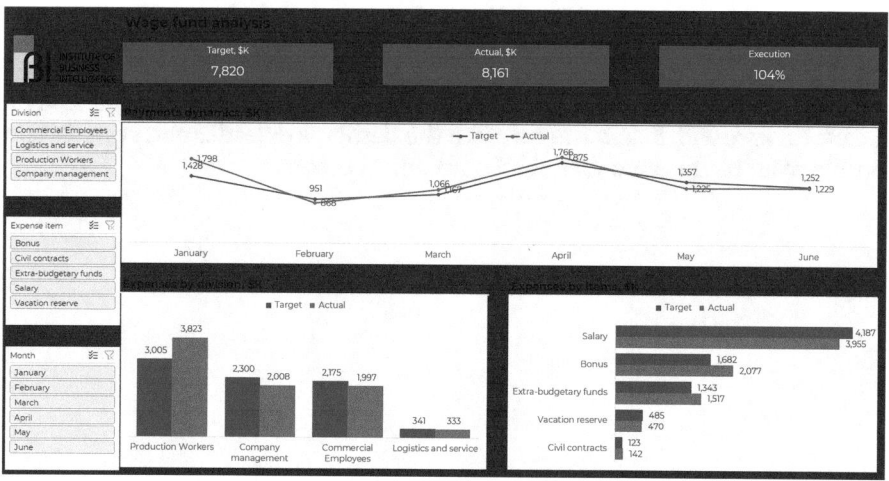

- • The logo became indistinguishable on a dark background.
- • The black font of the titles has also become invisible.

- Blue KPI cards do not fit into the new design.
- The background of the diagrams remained light – this does not correspond to the dark theme.
- The design of the slicers also does not correspond to the dark theme of the dashboard.

And these are not all the disadvantages. The colors of the columns and lines, as well as the data labels and the font color of the inscriptions remained the same as in the light theme. But we are making a dark dashboard: as soon as we change the background of the diagrams, they will also be lost.

Dashboard Title and Logo

I called the sheet with a copy of the light branded dashboard "The dashboard Dark": I will carry out the transformation on it.

The first thing that attracts attention is the "faded" logo. Because of the new background, it became indistinguishable. We will take an option from a dark theme in a PowerPoint presentation. Just as we did in the last chapter.

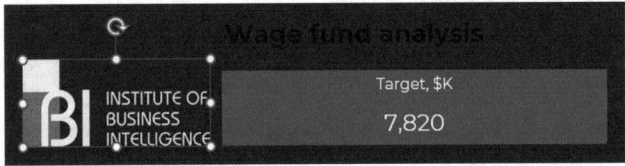

We also immediately see that the title "Wage fund analysis" has been lost on a dark background. Select the cell with the dashboard header and select white on the main menu tab under the Font Color button.

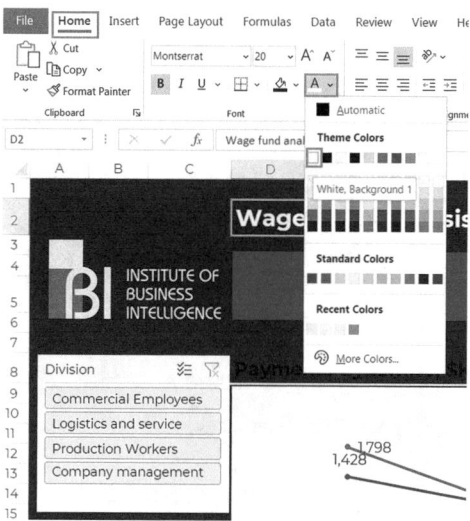

No operations with checking the font or changing its size are required here: we make a dark version on the dashboard with the "branded" font already configured.

Chart Titles

They have also lost their visibility on a dark background, which means that we are changing the font color of these elements as well. So that they do not "take away" the brightness from the title of the dashboard, we will choose the lightest shade of brown from the palette (90% lighter than the background).

The line under the titles. This element also remained light, and this also needs to be changed: in the dark version, semantic blocks should also be clearly separated from each other. First we will adjust the width of this line.

- Select the cell with the chart title.

- On the "Home" menu tab, click the "Borders" button.

- In the drop-down menu, select "Line style" ➤ the widest line.

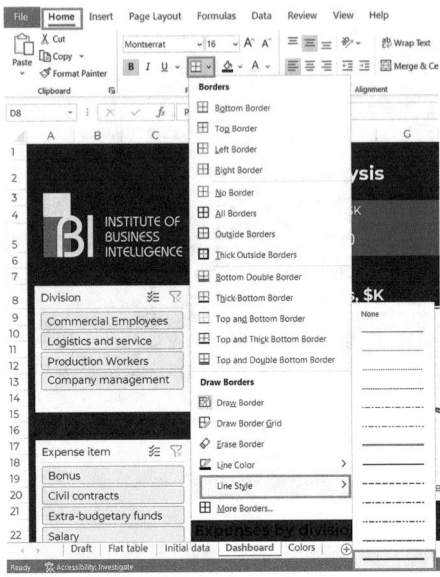

Next, in the same menu, click "Line color": from the suggested options, you need to choose a shade a tone lighter than our background.

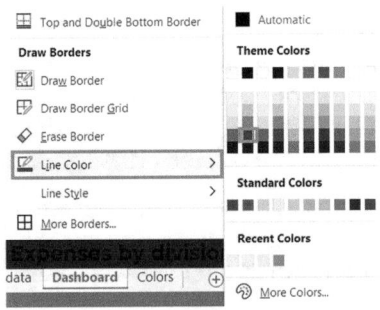

Do not be afraid of the "pencil" that has appeared: it draws exclusively straight lines. We can only use it to paint over the line under the chart heading. You can exit the drawing mode using the Esc key.

We set the same formatting for the titles of the other charts. You can repeat the preceding points for each, or you can select a formatted cell and use the "Format Painter" button to repeat this format for other titles.

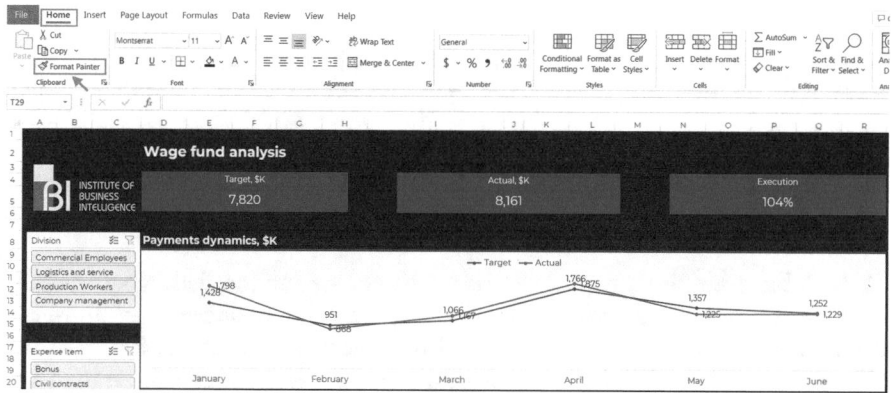

New Color for Cards

They catch the eye first of all – the saturated blue color does not harmonize with the new theme of our dashboard. If I recommended using a light gray or blue color of the cards for a white background, then here, according to this principle, I will take a shade of branded brown, only 25% lighter. It will turn out that the cards are "pressed" into the background.

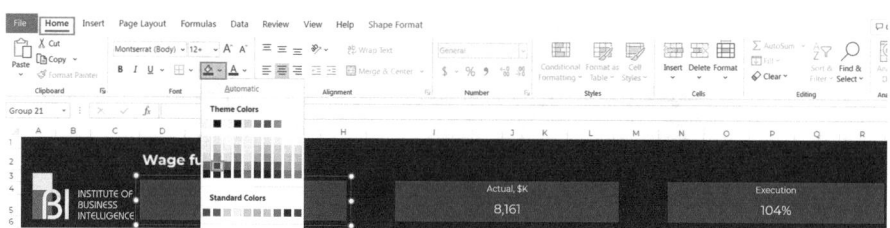

Setting Up a Graph

The background of the graph. It still remains in the light theme, and we need to bring it in line with the color of the entire dashboard. To do this, select the plot area and on the main menu tab, click "Fill" color to select the darkest shade of brown – the same one that we used for the dashboard background.

A line on the graph. Now a new task – against such a background, the blue line graphics faded. Let's change its color to green. To do this, select the plotting area and open the menu with the right mouse button. Here you need to use both buttons: "Fill" and "Outline."

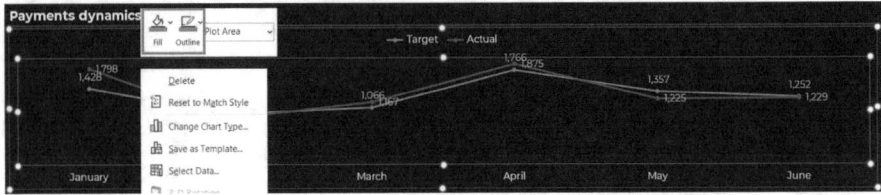

Data labels on the graph. We select the color for the data labels already in the menu ribbon. Usually I make them in the color of lines or columns, but on a dark background, this option will read poorly. Therefore, for both the "Target" and "Actual" labels, I choose a shade several tones lighter than the corresponding graph lines. It's not difficult with red – the right color is in the available palette.

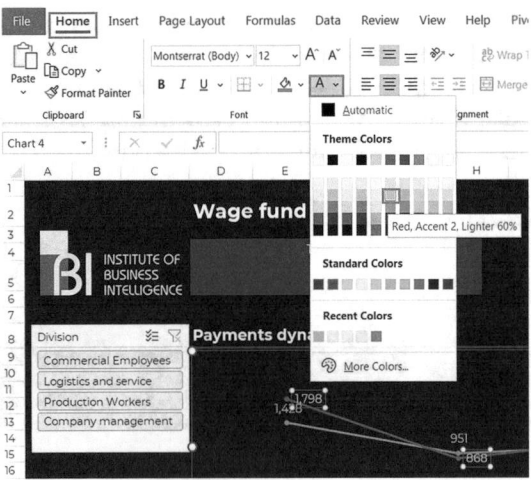

But for green shades in the layout of colors will not work – too "neon." Therefore, we will choose the color for the labels for this line ourselves. That's how I do it.

- I select data labels for the "Target" line.
- I choose the base green color for the labels (the same as on the line).

- I click "Font color" again in the menu ➤ "More colors."
- In the window that opens, I choose a shade lighter than the line.

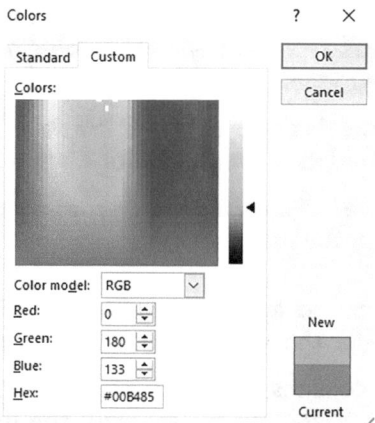

The color of the inscriptions on the axis and in the legend. After changing the background of the diagrams, the category labels "faded," became difficult to distinguish. And along with them – the names of the parameters in the legends. To correct this, we will change the font color to the lightest shade of brown for both category names and captions in the legend.

Charts by Template

The graph is fully customized and adapted for the dark theme of the dashboard. We use it as a template so as not to repeat all the steps in the rest of the diagrams.

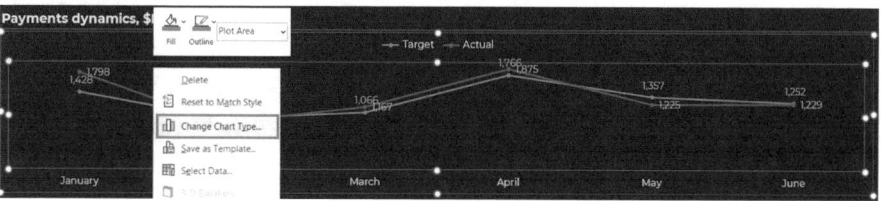

1. Select the graph and select "Save as template" in the context menu.

2. Save the template in the suggested folder (do not change it), giving it a clear name.

3. Select the "Expenses by divisions" chart and select the "Change chart type" option in the context menu or on the "Designer" tab.

4. In the window that opens, in the Templates folder, select the customized dark template with a graph.

5. Again, use the "Change chart type" option and select the histogram in the menu.

Not everything turns out as it should automatically. If the X-axis labels are placed at an angle, reduce the font size – they should only be horizontal.

Also on the "Target" columns, the green color has changed to yellow. We return what we need: select any column with the mouse (all the "Target" columns should be highlighted) and select the green fill.

Repeat the steps using the template for the third diagram "Expenses by items." This is how the diagrams should look like in the end:

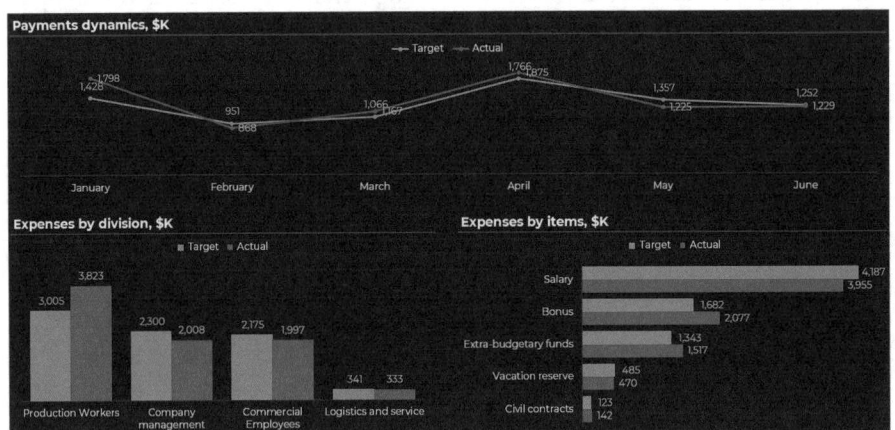

Setting Up Slicers in a Dark Theme

Interactive slicers remained the only unformed element on the dashboard. There is no suitable one among the ready-made styles for them, so we will configure them ourselves.

Let's take the style with red buttons from the "Dark" section as a basis – we will bring it to the desired result. In the list of styles, it is called "Rose."

To change it, select any slice on the dashboard. On the menu tab "Slicer" (or "Options"), right-click the desired style ➤ "Duplicate."

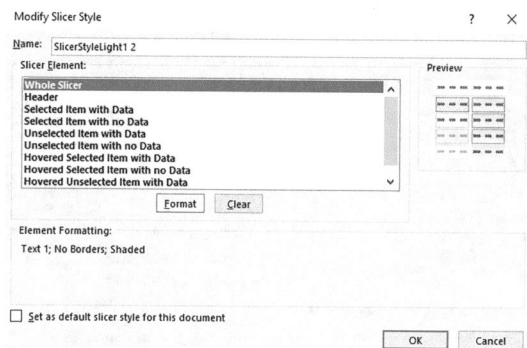

In the window that opens, in the section "Slice Element" – about ten items. Most of them are needed to customize the appearance of the slicer buttons in different states: when selecting, when hovering the mouse, and so on.

In order for everything to work as it should, we will have to configure each of these items separately. We will do this according to the same scheme: select the desired item, click the "Format" button, and add the necessary settings in a new window.

Whole slicer. On the "Border" tab, select the "No" option. On the Fill tab, select the darkest shade of brown – the same one that we used for the dashboard background. Click "OK."

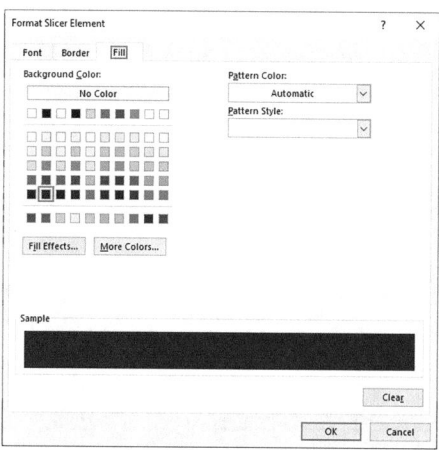

Header. This is the second item in the list of slice elements. On the "Font" tab, set "Montserrat (Main text)," and also adjust the color. Here we will set the lightest shade of brown – the same one that we chose for other headers.

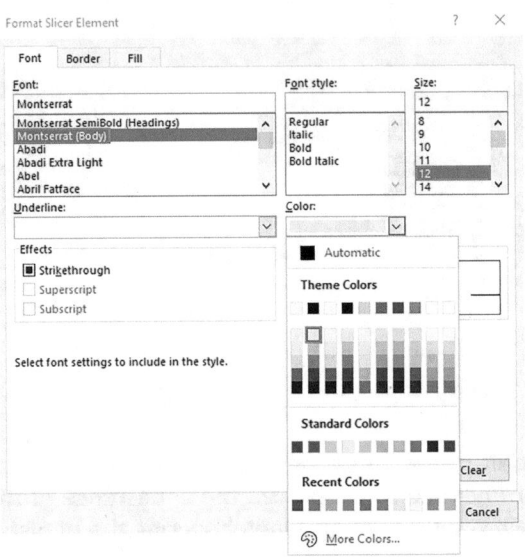

Selected Item with Data. Here we configure the type of active buttons, that is, pressed for filtering. By default, the Fill tab has a color that will suit us – we leave it unchanged.

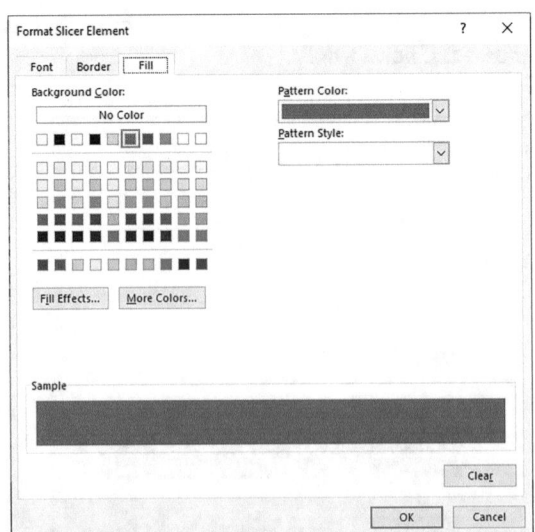

Selected Item with no Data. These are the pressed buttons, under which there is no data. We are satisfied with the default settings – we leave it as it is.

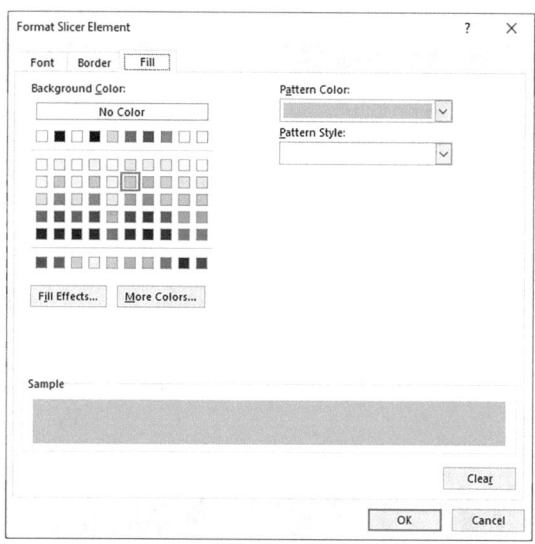

Unselected Item with Data. This is the color setting of inactive buttons when using a slicer. By default, when filtering, they will turn gray: we will change the color of this fill to brown – we will choose from the shades between the darkest and the lightest.

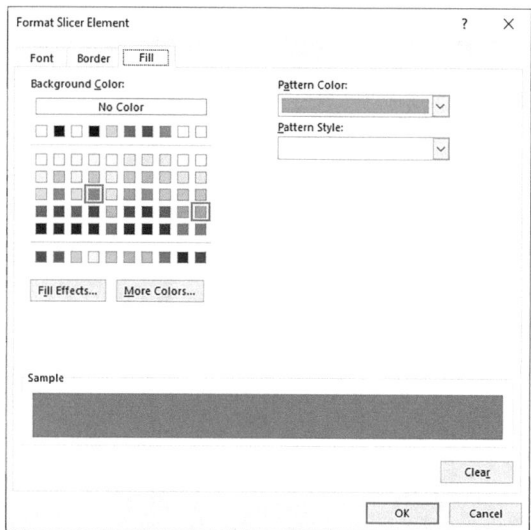

Unselected Item with no Data. Here we take the light brown color of the fill. It will be applied to inactive buttons with no data under them.

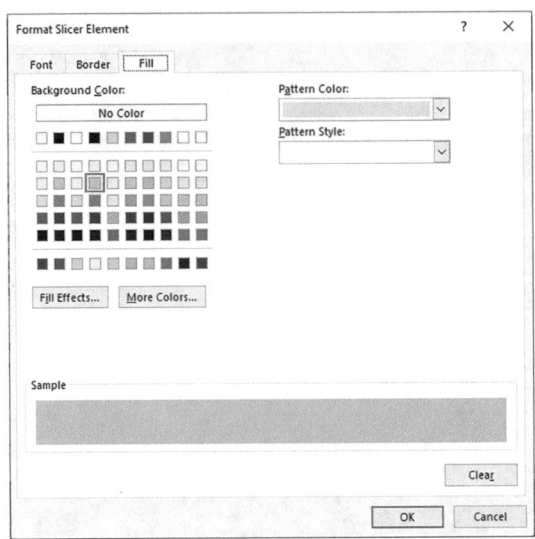

In the section "Slicer Element," we have four items left. All of them are needed to adjust the appearance of the buttons when hovering the mouse.

- Hovered Selected Item with Data
- Hovered Selected Item with no Data
- Hovered Unselected Item with Data
- Hovered Unselected Item with no Data

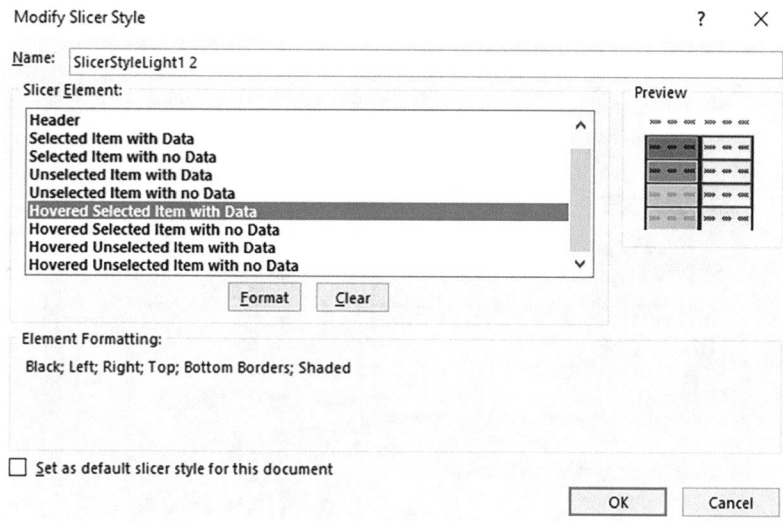

For these remaining items, we will set the same settings

- On the Fill tab, we will select a shade a tone lighter than the dashboard background.

- On the Font tab, we will set the light brown color and the Montserrat font.

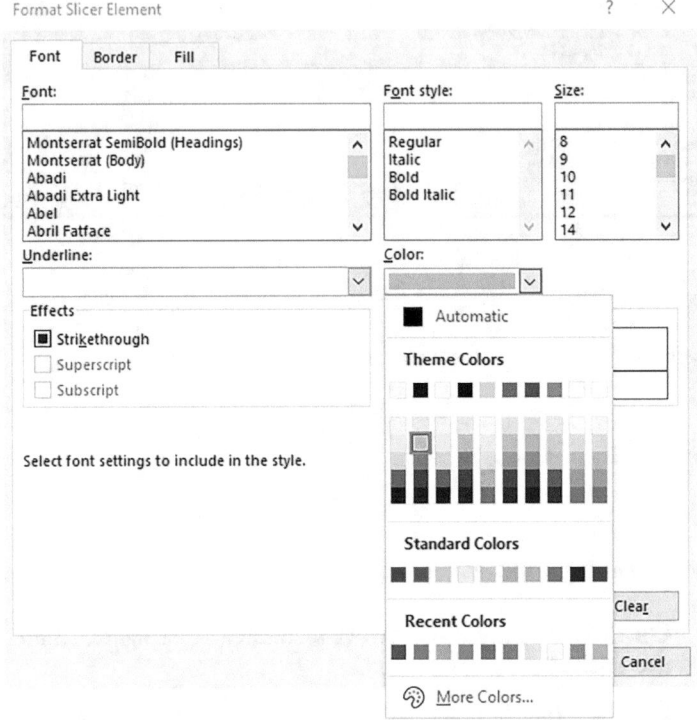

All elements are configured. In the "Modify slicer style" window, click "OK." The resulting style is applied to the rest of the slicers on our dashboard.

The dashboard "Wage fund analysis" in the dark theme of corporate identity is ready!

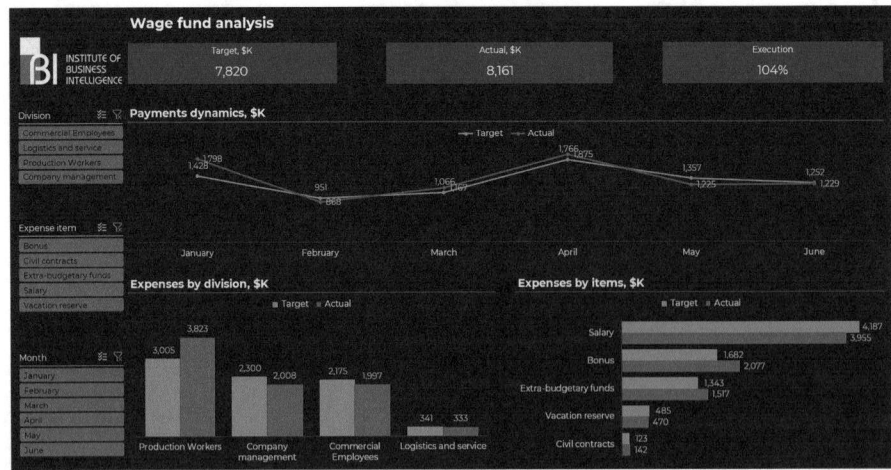

Summary

To create a professional dashboard in a dark theme, I had to pay special attention to the contrast of the elements. Somewhere we added it, and somewhere, on the contrary, we removed it.

- The headers of the dashboard and charts were made light so that they were clearly visible on a dark background.

- The lines under the headers were reduced in brightness and selected a shade a tone lighter than the background of the dashboard.

- The blue cards did not fit into the new design – we also made them a tone lighter than the background of the dashboard.

- We changed the excessively bright background of the graph: for this we used the same fill as for the dashboard background.

- On the new background, the blue line of the graph faded: we changed its color to green.

- We have selected new colors for data labels. Green from the palette did not fit, so we picked it up manually.

- Category captions have been made light so that they are visible on a new dark background.

A separate story is the slicers: I had to tinker with them, adjust the color of the buttons for each state. I admit, I don't remember all these combinations by heart myself. This once again showed that Excel can do anything. The one who knows its capabilities gets a lot of freedom of action. And it can do its job much faster.

Data Visualization Rules

We are done with the practice – now let's move on to the theoretical part. Usually people do the opposite, but I use this approach on purpose. Firstly, so as not to limit you to theory right away. Secondly, because there are exceptions in every visualization rule – I will also show their examples in this part.

These rules are relevant not only for Excel. They will help out when working in any BI platform if you want to accurately and quickly convey the essence of the data to the user.

The information in this part is a clear base for beginners, which will help accurately display the meaning of the data and increase the value of analytical work for any customer.

© Alex Kolokolov 2023
A. Kolokolov, *Make Your Data Speak*,
https://doi.org/10.1007/978-1-4842-8942-6_6

6.1 Types of Data Analysis

Some employees somehow miraculously manage to make simple and understandable presentations that the management likes. And others suffer; they redo it five times, but the boss still remains dissatisfied. And even if they hire a designer, it doesn't get any easier.

I will share my personal experience. At the beginning of my professional career, I was engaged in the commercialization of scientific projects, looking for investors. And there was a strict format of investment presentations: on ten slides, it is necessary to state the essence of the project, show market analysis, financial model. I somehow managed to clearly visualize it by myself.

Later, when I had to make the first dashboard, I hired a designer to "make it beautiful." It turned out bright, original, but unclear. Then I read the book "Say it with charts" by Gene Zelazny and realized that I followed the rules of visualization intuitively. But I couldn't clearly explain to other people why it was necessary to build such a diagram.

Over time, with experience, this technique appeared. With its help, I established communication with customers, began to ask them the right questions, understood what they wanted to see, and what decisions to make. And then I could clearly set the task for my analysts and developers in order to get the desired result the first time.

What should the diagram show? What conclusion should it lead to? It depends on the type of data analysis, that is, what your data is in meaning.

Basic Types of Analysis

Rating

Quantitative comparison based on the principle of "more-less": who is in the first place, who is in the top 5 and by what margin, and who is how far behind.

Visualization: columns, horizontal and vertical

Dynamics

Changes in indicators over time: revenue growth or decline, the number of staff by month, and so on.

Visualization: usually lines on the timeline

Structure

Analysis of the composition of the whole, where the emphasis is not on quantity, but on fractions of 100%. Helps to understand which segments and categories contribute to the achievement of the indicator.

Visualization: figures divided into sectors

Distribution

The distribution of a quantity on a quantitative scale, constant or interval. For example, segmentation of customers by age or check amount.

Visualization: scatterplots, histograms

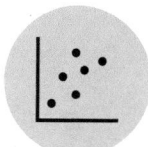

Relationships

Displays dependencies between indicators, as well as logical relationships between categories. For example, how financial flows move in the holding.

Visualization: graphs and other complex diagrams

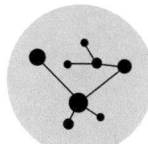

The last two types of analysis, distribution, and relationships, are rarely used in management reporting. They belong in statistical analysis, where you need to look for correlations and build complex graphs. To present a report at a meeting, you need simple and understandable visualizations that accurately convey the meaning of the data.

There are more complex types of analysis (factor analysis, cohort analysis) – combinations of several basic ones. There are advanced visualizations for them: some can also be built in Excel, while others will already require full-fledged BI systems. But first we need to deal with the database and avoid mistakes at this level.

Typical Errors

Someone will say that the following examples are some kind of childish mistakes. But on projects in large corporations, I regularly see how adult people make them in their reports. Because "it is accepted" and "we are already used to it." Or because once the director liked such a schedule, and now everyone is building it about and without.

Pseudo Dynamics

Here we look at the profit chart for managers. But what conclusion can be drawn from this line? It seems as if it depicts periods of growth and decline. Williams had a peak, then Wilson had a decline, then Johnson had a peak again, and then a smooth decline.

If you need to compare the profit by employees, then first of all it is important to see who is in first place, in the top three, the top five, and by what margin from the rest. The same is true for laggards.

But sorting managers alphabetically does not help to understand who is in what place. And the line of the graph is torn off from the surnames: first we see the maximum value, and then we look with our eyes to whom it refers.

On the X-axis, we sort categories only for dynamics and statistical charts. In other cases, we order the values from larger to smaller along the Y axis.

The line graph is used to track the change in the indicator over time, but not to compare categories on the X-axis with each other.

To compare employees by the amount of profit earned, a column chart is perfect. It is immediately clear: there are two leaders, there is manager Brown in third place, and after him — two are competing for fourth at once. The second half of the department is lagging behind.

It is clear that there is not enough data for serious conclusions about labor productivity. You need to compare the fact with the plan, the volume of the customer base, and other indicators. But the fact itself on such a column chart is shown as clearly as possible.

Inappropriate Funnel

Many people believe that the funnel diagram is a symbol of an advanced CRM system, great sales analytics. Therefore, when they find it in MS Excel or Power BI, they try to use it at every opportunity. It often turns out to be out of place, and without explanations, it is impossible to understand what the author wanted to convey to us.

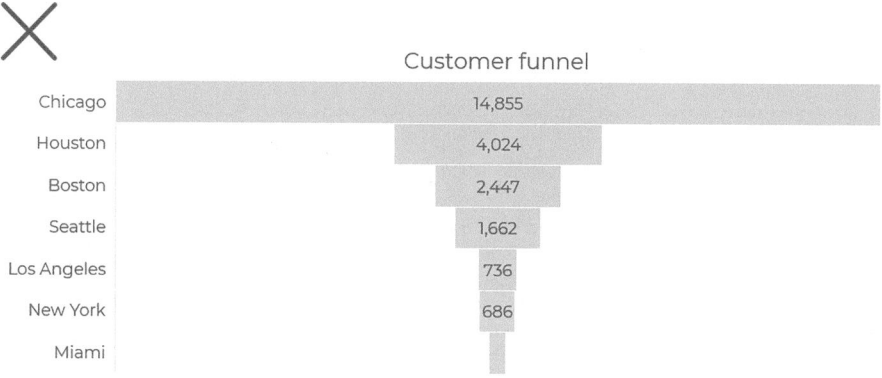

Here, with the help of a funnel, they tried to compare branches by the number of customers. But from the outside it looks like this:

- 4024 clients have moved from Houston to the Boston branch.

- Of these, 2447 customers moved to a branch in Seattle.

- And so the customers moved through the cities until they ran out. :)

Of course, the author wanted to show something else: which branch has more customers. That is, a quantitative comparison on a scale is a rating. To do this, you do not need to shift the columns to the middle: we build them from zero on a regular bar chart.

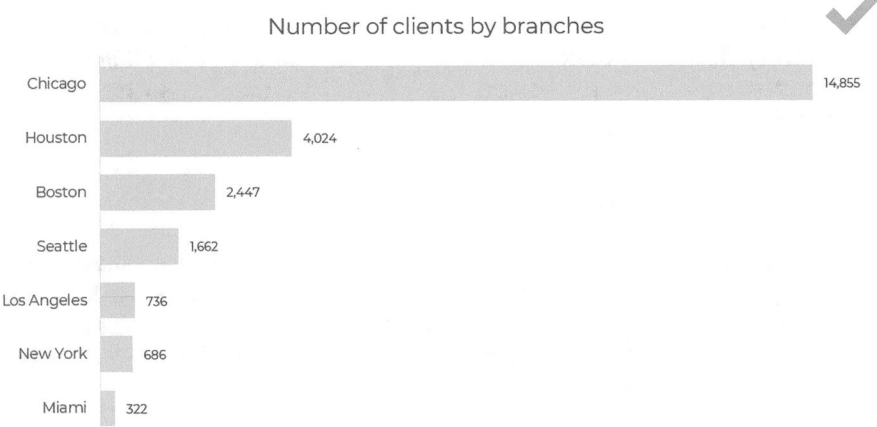

Number of clients by branches

Chicago	14,855
Houston	4,024
Boston	2,447
Seattle	1,662
Los Angeles	736
New York	686
Miami	322

"But in the funnel, the values just go from more to less!" the students say. "What's wrong with the rating?" I'll explain. The funnel is not needed to compare indicators.

The funnel diagram is a special case of dynamics. And it is needed solely to visualize a step-by-step process in which at each stage there is a dropout.

For example, it can be used to show how the number of clients changed from the first call of the manager to the conclusion of the contract. You see that after the presentation, few customers reach the demo version, but then there is a high conversion into a proposal and a contract.

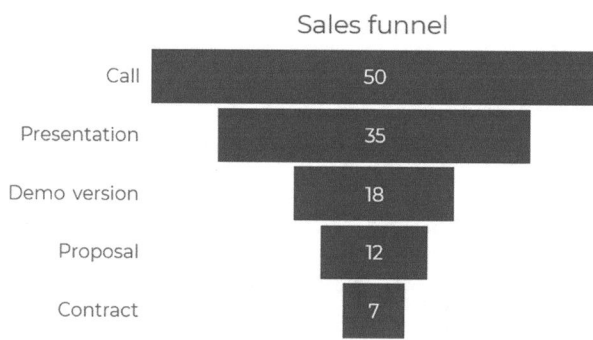

Not All Pies Are Equally Useful

Pie charts are used most often for other purposes. Let's take a look at this visualization of the project portfolio.

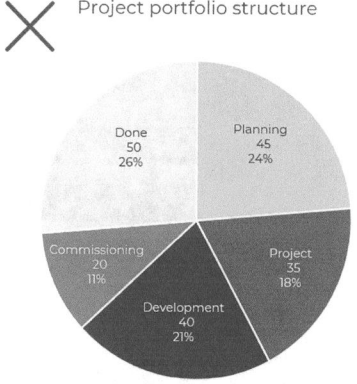

Everything seems to be done properly. A pie chart is selected for the structure, there are only five sectors on it, there are data labels, and categories titles fit horizontally. But no matter who I show it to, everyone doesn't like it: someone suggests taking the labels outside the circle, someone says that you need to change the colors... But the problem is different.

The purpose of the diagram is to show a state by project stages at the moment. Here we do not sort the stages from more to less, they follow each other: first planning, then design, then development, and so on. In fact, this is a chronological sequence, only not by calendar periods, but by stages of the project.

This is not a structure, but a dynamic – we always display it from left to right along the horizontal axis. You can immediately see that 50 projects have been completed, 20 are on commissioning, 40 are at the development stage, and so on. Data labels with percentages, which are common for a pie chart, are not needed here and only confuse.

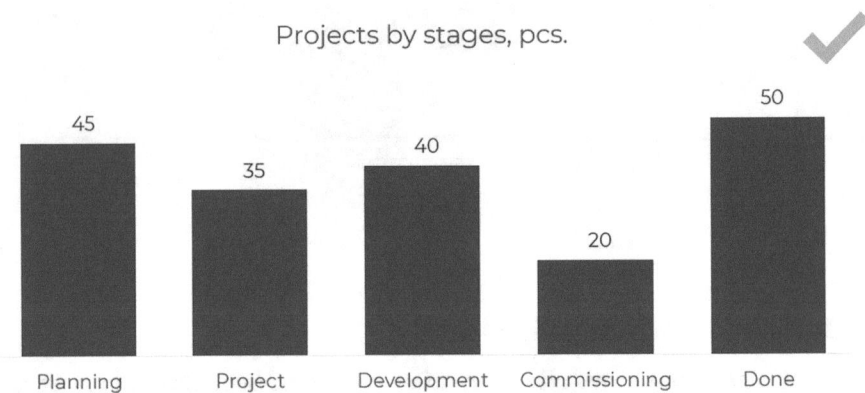

Once again, I will draw attention to the horizontal timeline. It may seem to some that a bar chart will work better, especially with longer labels.

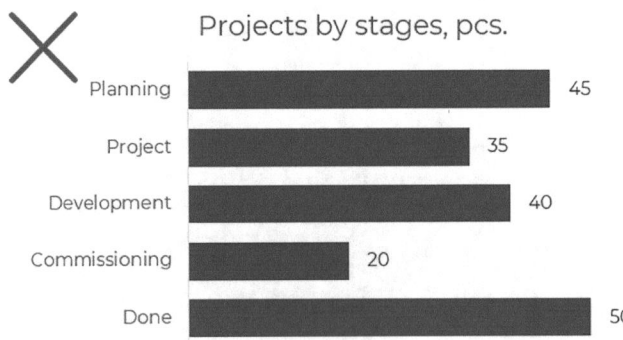

This will be, if not fatal, but a mistake. By default, the eye perceives horizontal columns as a ranked list, and we need an emphasis on dynamics.

There is another example of inappropriate use of pie in which there was an attempt to show which age groups our customer base consists of.

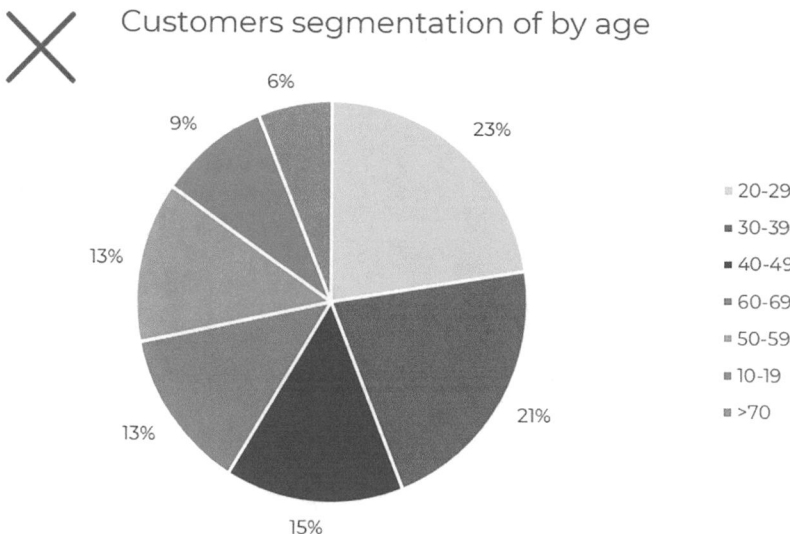

The sectors are sorted exactly as needed on the pie chart – from larger to smaller. But segments are age distribution intervals. That is, again we are dealing with a horizontal scale from minimum to maximum.

The most obvious option is to show the size of the segments in columns. They no longer need to be colored, as it was with the sectors, and there is no need for a legend either: the age intervals are signed on the X-axis.

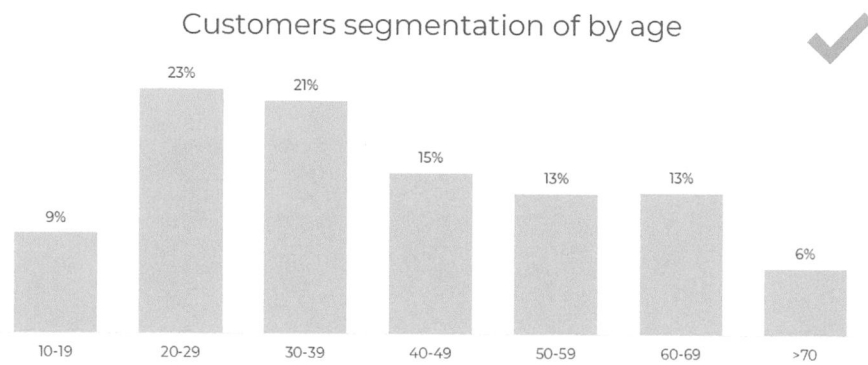

You may wonder why correcting all the errors leads to charts with horizontal or vertical columns. But this is really the most functional way of visualization.

Any kind of diagram should be chosen consciously. First, we determine what the data is in terms of meaning and what kind of data analysis it is. And only then we choose the appropriate diagram. This is not a creative, but an exclusively logical process.

There are several variants of relevant diagrams for each type of analysis. The instructions in the following sections of this chapter will teach you how to choose correctly in each case. You can master the skill of data visualization without being a designer.

Summary

5 Types of Data Analysis

1. Rating – for quantitative comparison of indicators

2. Dynamics – to understand changes over time

3. Structure – for the analysis of the composition of the whole, a fraction of 100%

4. Relationships – to evaluate quantitative and logical relationships between several parameters

5. Distribution – for segmentation on an interval or constant scale

In 90% of cases, only three of them are used in corporate reporting: rating, dynamics, and structure.

To avoid mistakes, you must first determine the type of data analysis, and then select the appropriate chart.

6.2 How to Choose Charts

There are suitable and inappropriate diagrams for each type of analysis. For example, it is not a good idea to display the monthly dynamics on a circular one – we draw a line for the dynamics although you could also see bar charts on the timeline. Then the question arises: which option is better?

First of all, it depends on the amount of data: five columns is one thing, and 15 is different. I will tell you about the recommended rules, but they cannot be considered a strict algorithm. You will constantly encounter exceptions to the rules, because the choice of a diagram depends on other nuances. For example, it depends on the length of the category labels and the space allocated to the diagram on the screen.

Additionally the choice of the chart depends on how many data series you want to put on it: just actual indicators or their comparison with the plan, last year, industry benchmark. In this part, we'll start with a simple one: how to choose charts for ranking, dynamics, and structure with a single row of data.

Charts for Rating

To visualize a quantitative comparison, we use a figure called a column, which can have a horizontal or vertical orientation.

But in MS Office, these are two separate charts: the vertical one is called a column chart, and the horizontal one is called a bar chart. Both are suitable for a rating with a single row of data, but are not always interchangeable. Let's figure out in which cases it is not necessary to use everyone's favorite column chart.

Choose by the Number of Categories

When there are many columns on the column chart, the category labels do not fit horizontally, and it is difficult to determine which column the data label belongs to at an angle.

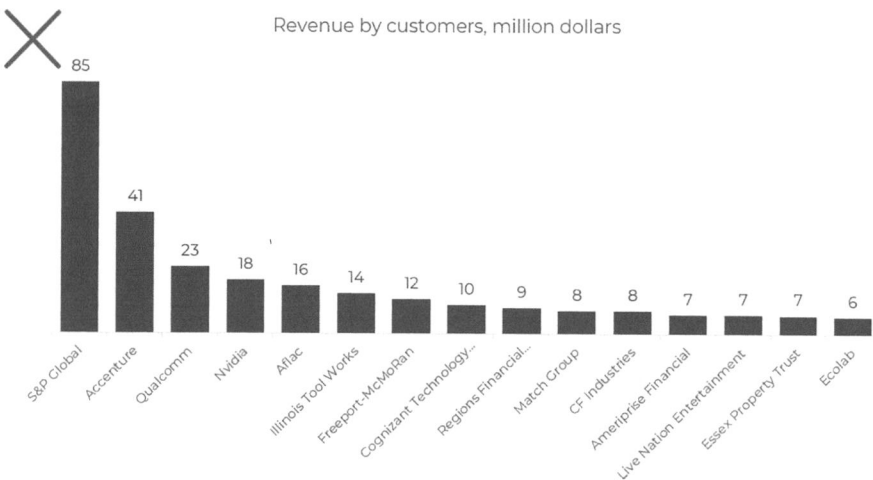

To solve this problem, the diagram needs to be "flipped," and then the category labels can be easily read horizontally from left to right. As a rule, if you have up to ten0 columns, the labels are placed on the vertical bar chart, and over ten – on the horizontal.

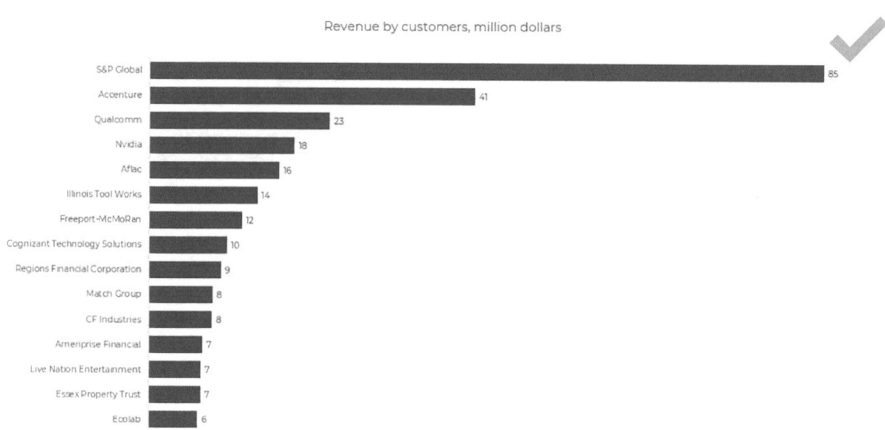

I remind you that ten categories are an approximate guide, you need to take into account other nuances.

Consider the Length of the Labels

In this example, there are only eight types of services, but they have very long names that are again at an angle.

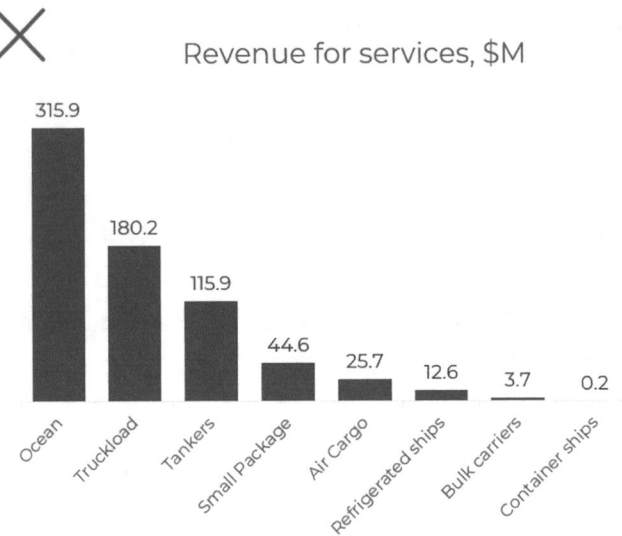

Here again, a bar chart will help out although I recommend choosing abbreviated names for long types of services or cost items. The space should be used for visualization, not redundant text.

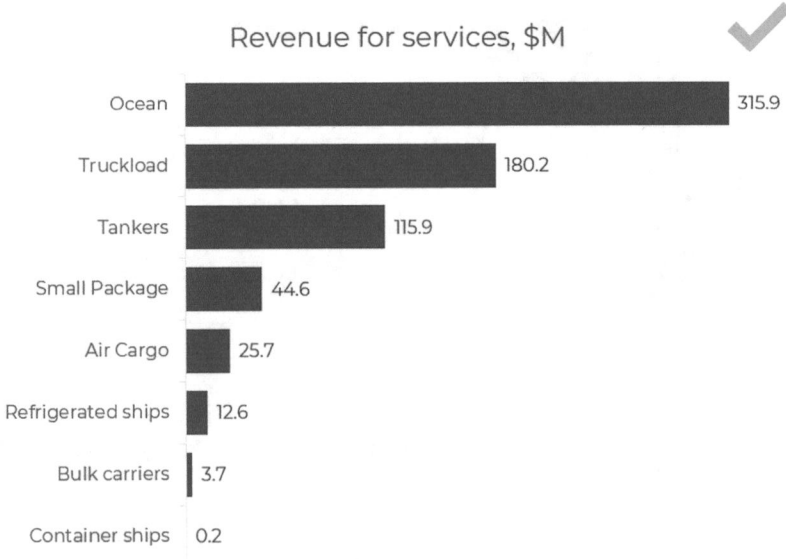

Estimate the Available Space on the Screen

Revenue from seven branches on the column chart will be visible if there is space for it at least half the width of the dashboard or slide screen. But here we have about a quarter, and the labels again do not fit. And there is no such problem for a bar chart – it performs its function well.

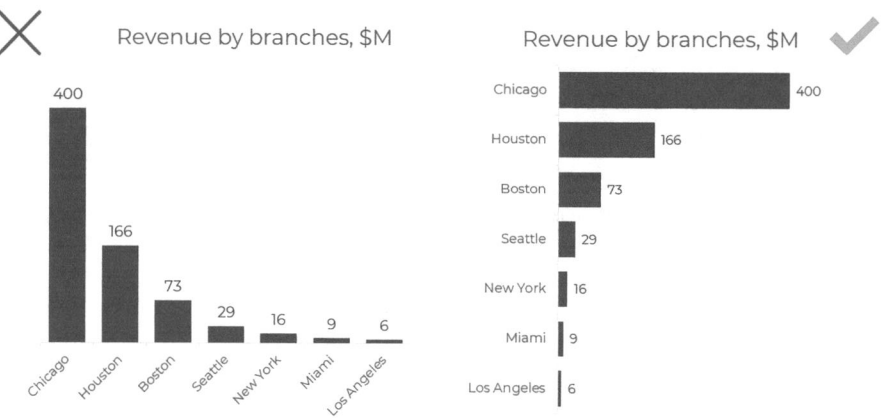

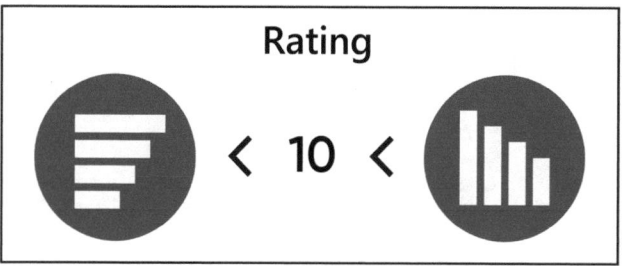

Charts for Dynamics

To show how one indicator has changed over time, a line graph or a column chart is better suited. In some cases, you can also use a graph with areas – I will not focus on it yet.

Let's figure out when to choose a chart with vertical columns, and when a line graph. The criterion is still the same – the number of data points; only here it increases to 12.

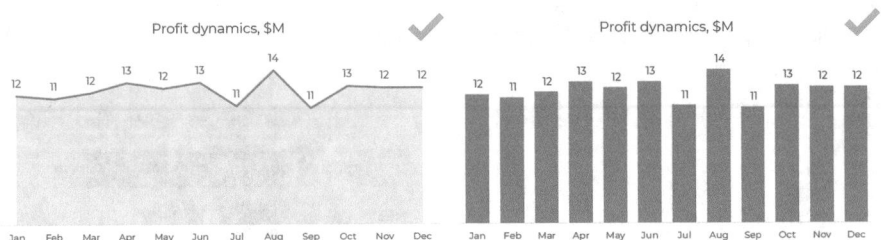

12 periods will be visible both on the chart and on the column chart, because the names of the months are shortened. But if the months of the next year also appear on the dashboard, there will be an overload.

In this case, the chart will better focus attention on the trend line and get rid of unnecessary visual "noise."

Less Than Six Periods

The reverse situation is when we have only a few data points, for example, four quarters of the year. In my lectures, I show two diagrams side by side: I warn that the data is the same, and I ask which one is the best at first glance. Most choose columns.

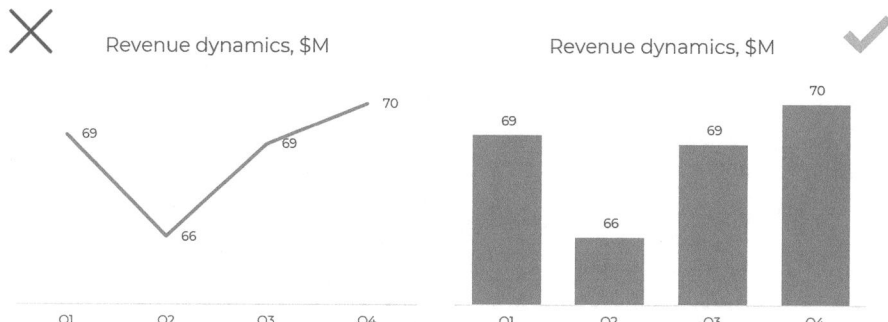

I agree with them: the line graph looks disconnected from the scale and somehow empty. Smooth lines between points are also confusing, because growth or fall are rarely so uniform. There is not enough data for a convincing trend line, there is no trust in it. And on the column chart, the emphasis is more on the amount for the quarter, but at the same time, the chronological order is preserved.

A Common Mistake with the Direction of the Scale

Sometimes mini-charts with dynamics are added on dashboards right under the KPI cards (or even instead of them), for example, for several years. And often bar charts are used for this.

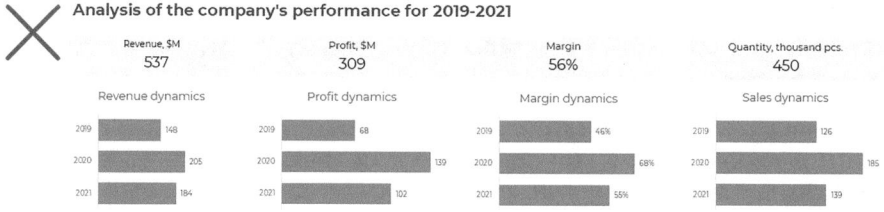

Don't make such a mistake! We have the timeline strictly on the X-axis, that is, horizontally. So, for such mini-charts, we choose mini column charts.

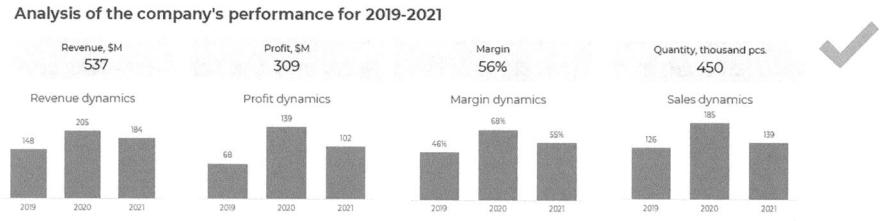

BRIEF CONCLUSIONS: VERTICAL COLUMNS OR LINES?

- If there are less than six periods, choose a column chart.
- If there are a maximum of 12 periods, both are suitable.
- If there are more than 12 periods, only a line chart.
- A bar chart has no place in displaying dynamics.

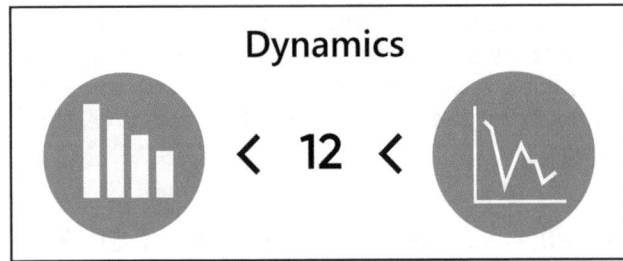

Diagrams for the Structure

To show which parts make up something whole, most often use a pie chart or its kind – a doughnut chart. There is no difference between them, so the choice is yours. But there is another way to display the structure – this is a treemap chart.

Both the pie chart and the trimap are divided into sectors – the same shares of the whole. In sectors, you can show, for example, the share of revenue in each city. Or to see the full structure of the company's expenses, if different items of expenses will be located in the sectors. Let's figure out when to divide a circle into parts, and when to divide a rectangle.

Six sectors – Maximum for a Pie Chart

Everything that will be said later about the pie chart is also relevant for the doughnut chart. Both of them display the whole, and the form imposes some restrictions on both. One of them is the allowed number of sectors. There should be no more than six sectors.

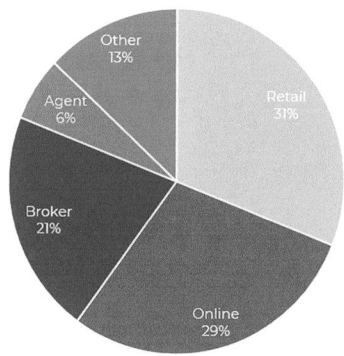

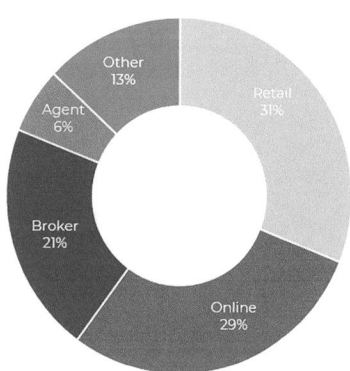

Otherwise, the sectors of the circle will become too small and it will be difficult to understand what the difference is between them. In addition, you will not be able to place labels in them. Taking them outside the diagram is a bad idea: this will confuse the user and create a pile of elements.

If there are from 7 to 10 sectors in your whole, it is better to choose a trimap diagram for visualization.

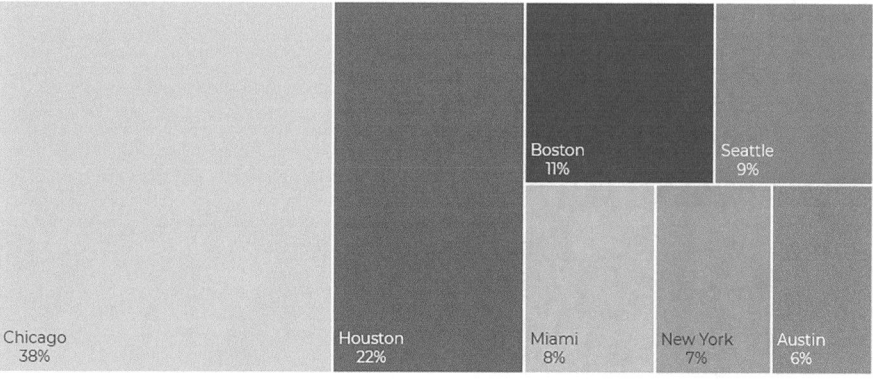

Long Labels? Give Up the Pie Chart

Category titles may not fit even with a small number of sectors. And this is also the reason for replacing the diagram. Treemap is larger in size, and therefore it is easier to place titles in its sectors that did not fit into a circle or doughnut.

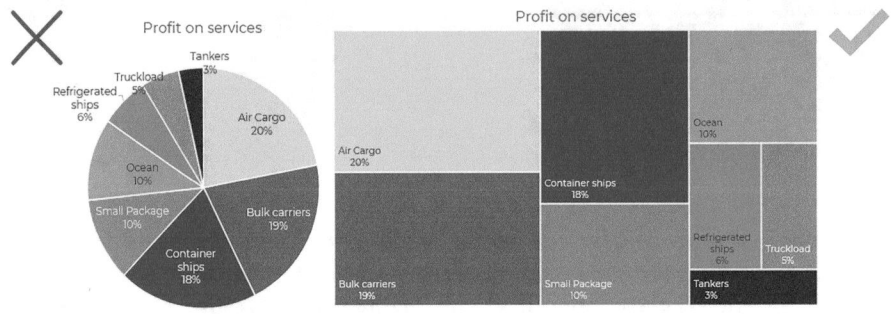

Tips for Designing Pie Charts

Pie charts have a lot of drawbacks, but sometimes this is the best visualization option. The convenience of perception and data analysis on a pie chart largely depends on its design.

- Place the first sector so that it starts at the top of the circle, similar to the 12:00 point on the dial. Sort the sectors from larger to smaller.

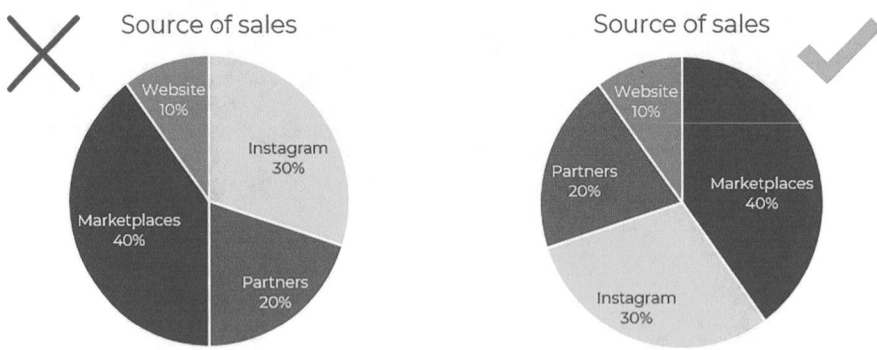

- Do not use a legend on a pie chart: do not force the user to run his eyes from the sectors to their names in the legend and match the colors of these elements. Place the labels inside the sectors.

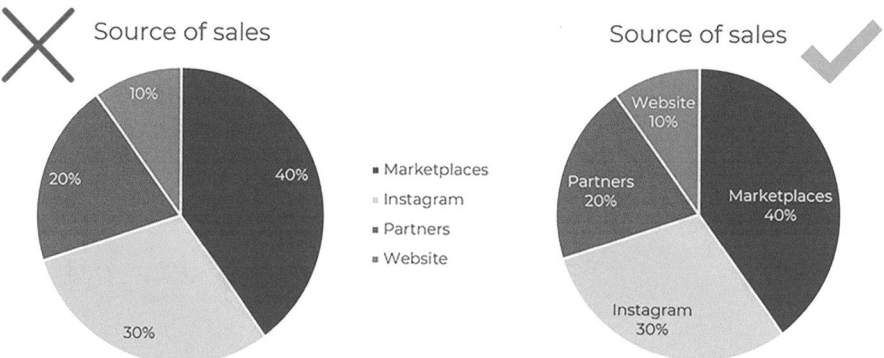

- It is better to put the labels of small elements outside the chart. Make sure that there are not too many of them, otherwise the labels will overlap each other.

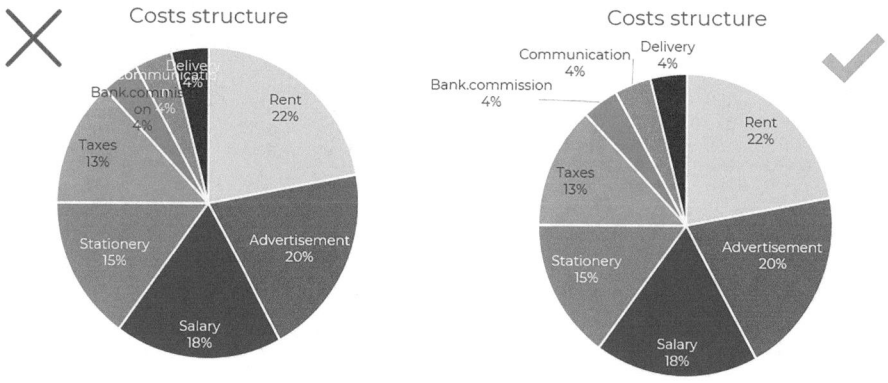

- Never. NEVER use 3D charts. For some reason, this is done most often with pie charts – to someone they seem more beautiful, to someone – more solid. But the 3D distorts the perception of data. The example clearly shows how the near "volume" sector of 29% looks larger than the sector of 31%.

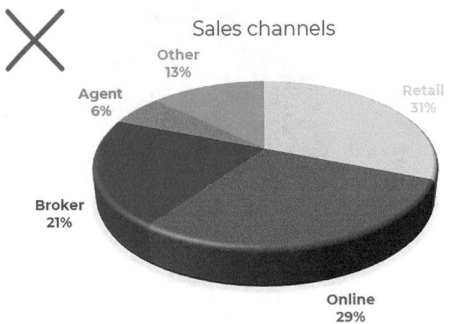

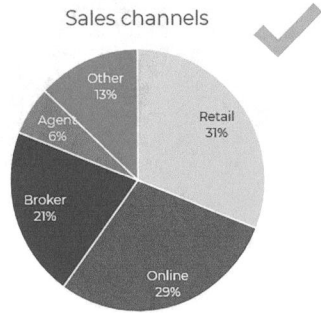

BRIEF CONCLUSIONS: HOW TO SHOW THE COMPOSITION OF THE WHOLE

- If there are no more than six individual parts in the structure, a pie chart or doughnut chart will do.

- If the labels do not fit into the sectors of the circle, take a treemap chart.

- Don't use a legend on a pie chart.

- If there are from 6 to 10 individual parts in the structure, the treemap chart is suitable.

- If the labels do not fit into the treemap sectors, use a bar chart as an extreme case.

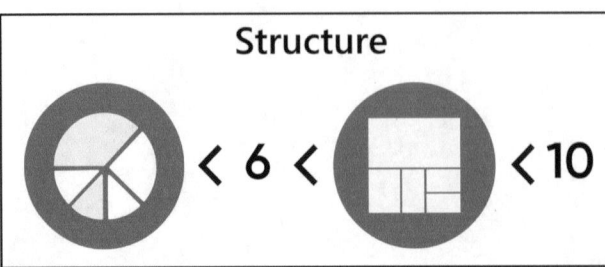

Summary

Even in three popular types of data analysis, you can get confused if you don't know the rules for choosing charts. Use a simple scheme.

- For rating: column or bar chart
- For dynamics: a line graph, a column chart, or an area chart
- For the structure: pie (doughnut) or treemap chart

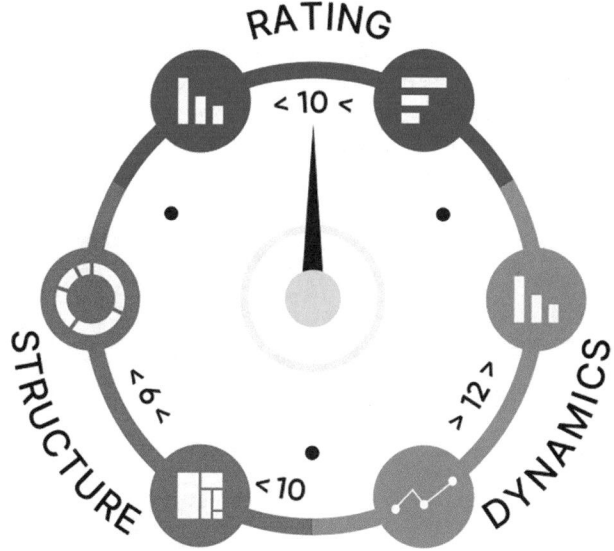

6.3 Life Hacks for Multiple Data Series

In the last part, we worked with one data series that showed only the actual value. But in real reports, you often have to deal with several rows, compare actual and targeted indicators in terms of rating, dynamics, or structure.

On the Wage fund analysis dashboard, we compared departments and expense items according to target and actual. To visualize these ratings, a column chart and a bar chart were used, although horizontal and vertical columns would be suitable for both cases on such data. For 4–5 categories, it will still be visual.

For dynamics, we built a linear chart, even though there were only six points. It may seem that this contradicts the rules described previously. But the data on the dashboard will be updated with new months, and one day there will be 12 of them. Besides, for the beginning, it was important for me to show you different charts, not just columns.

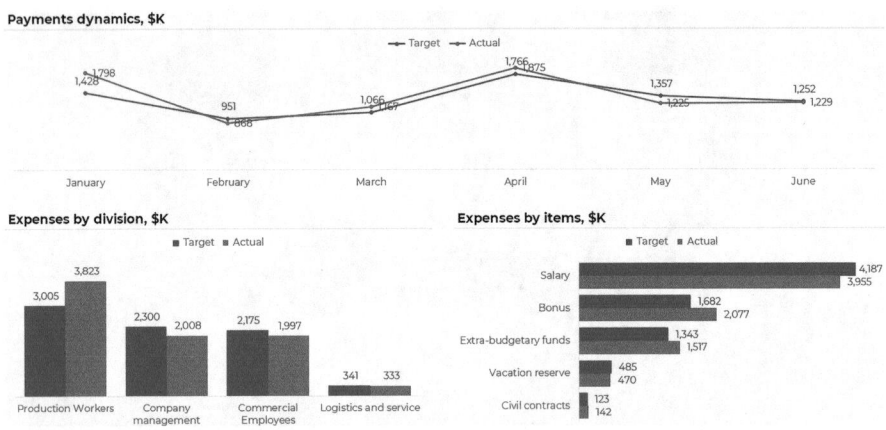

All of the preceding can be represented more compactly by combining series.

- For monthly dynamics – a combined chart with a line and columns.

- For the rating of divisions – a chart with markers that replace columns with targeted values.

- For the ranking by items, there is a chart with combined columns: the actual is here against the background of the target, as if a progress bar is striving for it.

I must say right away that this is not a reference visualization: Excel does not turn out as beautifully and visually as it could be in Power BI, for example. And anyway, our main focus has become on the actual, and the target is the background. Somewhere it may be good, and somewhere it may not be very good.

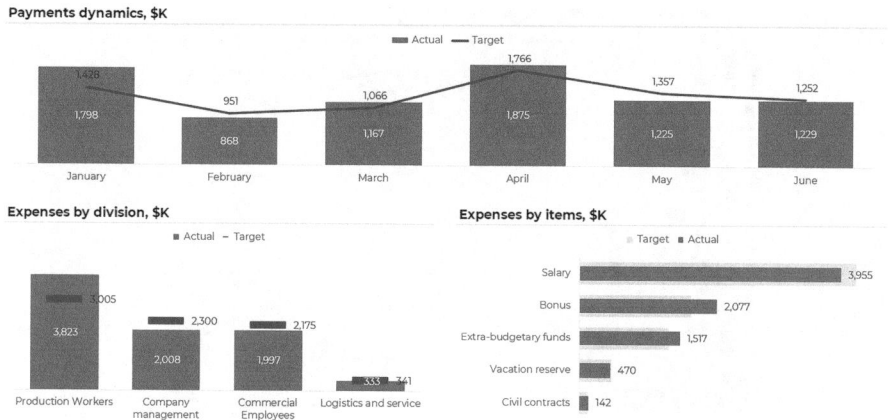

Let's figure out what tricks you can use to build these visualizations, as well as what pros and cons they have.

Combined Graph for Dynamics

All the preceding new visualizations have a common idea: the target is a mark that the actual crosses. To adjust this on the chart, select the line with the target and select "Change Series Chart Type..." in the context menu. If you don't have such an item, it means that you clicked on the chart area, and not on the series itself (line or column).

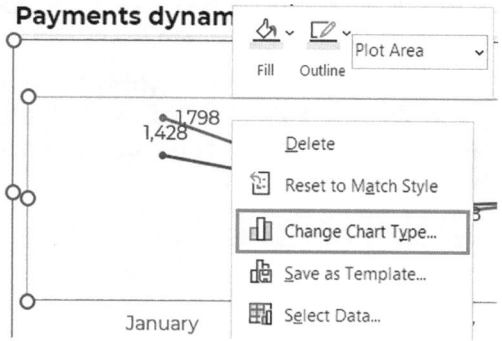

In the window that opens, we can select a separate chart type for each data series. In our case, for the "Actual" series, we select a Clustered Column Chart.

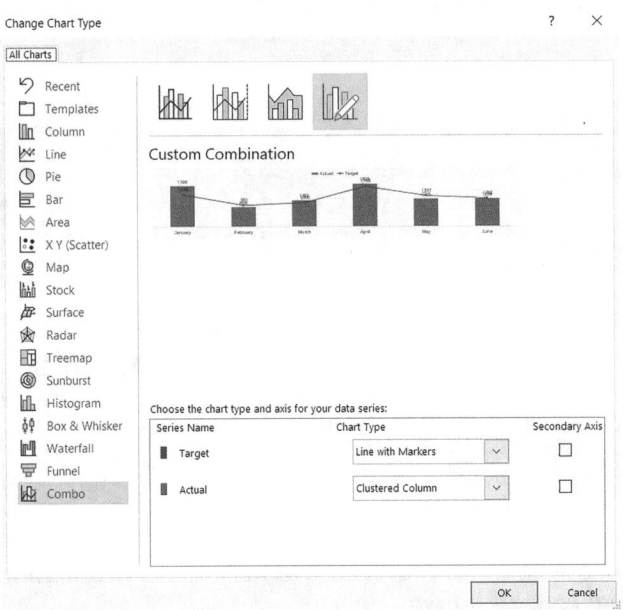

That's the whole trick. The advantage of the column is that the data labels can be placed in the center, and they will not overlap on the target, as it was before.

It remains to adapt the diagram according to the checklist from Part 3.4. First, we reduce the gap width from 150% to 50%. Then we increase the thickness of the "Target" line to 3 points so that it does not get lost against the background of large columns.

In this example, it was possible to leave the lines, everything is quite clear. But often you need to show three series: the target, the actual, and the actual of last year. And here such a technique helps out well. Just keep an eye on the accents: last year should not distract attention from the current one.

If you need to show more indicators for the dynamics, according to the rule, we choose a line graph. But when you need to see both the profit values for this year, and the target, and the values of last year, a combination of column chart and lines is better suited.

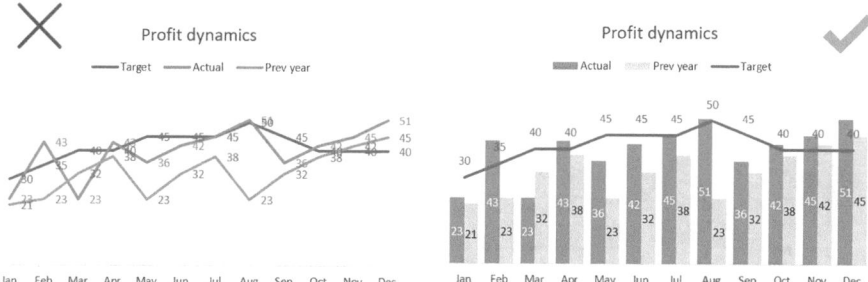

Sometimes it is necessary to place data of different scales on the chart. For example, revenue in thousands of dollars and a small average check in dollars. It will not be possible to simply reflect such data on the diagram: the line of the average check will "stick" to the X-axis. To prevent this from happening, the trend line along the middle check should be placed on the auxiliary axis.

For clarity, I colored the axes in the color of the lines so that it was clear which one refers to what. Just labels will not help here: they will become a poorly readable set of numbers mixed up. So the result will be better, although it will not be perfect.

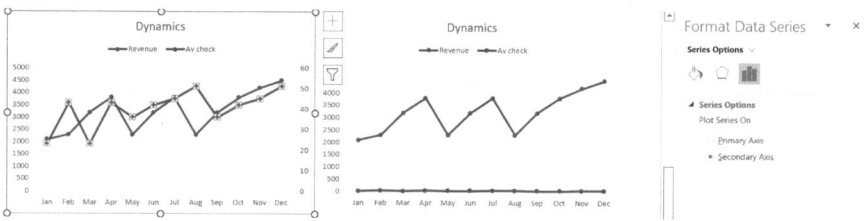

To strengthen the emphasis, we use a different technique – different figures for data of different order. We will display the revenue as a column, as a volume, total index, and we will leave the average check as a line, but as a calculated one.

Customize the visual representation according to the checklist – add labels, remove axes. But a problem arises here: the line of the middle check sticks to the X-axis again, although the series is built along the auxiliary axis. This is another feature of Excel, because of which we cannot simply remove the scale from the chart. But we can hide it: we will make the axes invisible by repainting them in the background color – we have it white. They are still there, but they no longer distract attention from the data.

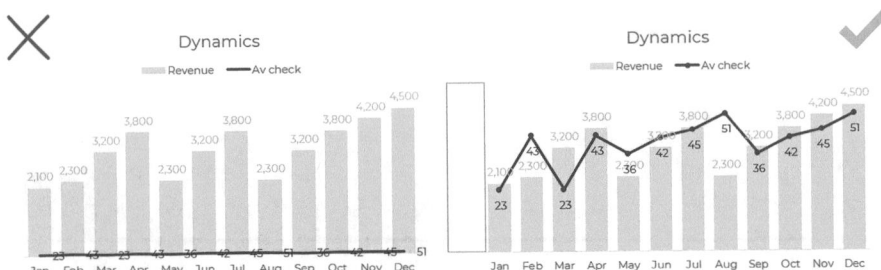

We need all these complexities and life hacks only to build such charts in Excel. In specialized software products, such as Power BI, such sophistication is not required.

Bullet Chart for the Target-Actual Rating

The Bullet chart is used to display target–actual values. The visual idea is that the strip with the actual value kind of shoots, flies out of the barrel with a target. The example on the left looks more like such a gun, and on the right is already a minimalistic version, where the target is a line, a bar that the actual reaches. Plus, conditional formatting is used here: if the goal is not achieved, then red, if completed, then green.

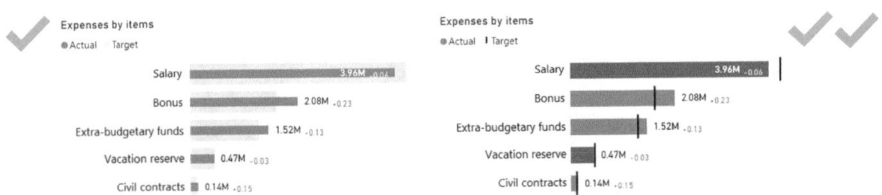

Both options are correct, but I recommend the second one. There is more space for the actual column on it, and the target is compactly displayed with a dash. But these are examples of a ready-made Bullet Chart built in Power BI. In Excel, we will have to conjure, and the result will not be 100% consistent with the standard.

Horizontal Bullet Chart with a combination of Series

To build a horizontal bullet chart, we first need to change the parameter for the "Target" data series and set the overlap of the series to 100%. All data labels are mixed up and therefore we delete them from the "Target" column.

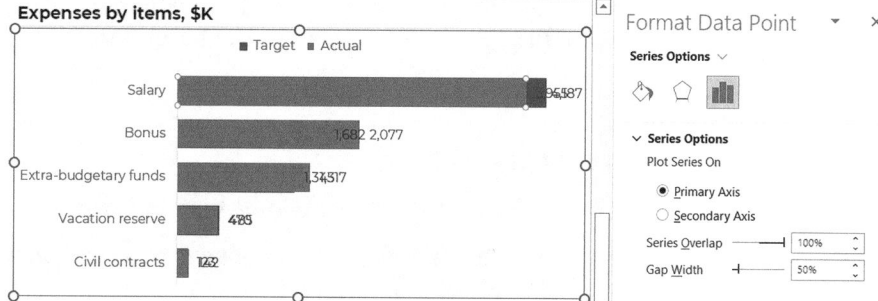

Then we switch the "Actual" to the auxiliary axis and set 0% for the Series Overlap, and 125% for the side gap. After setting the bit depth of the axis for the auxiliary series, we remove it from the diagram.

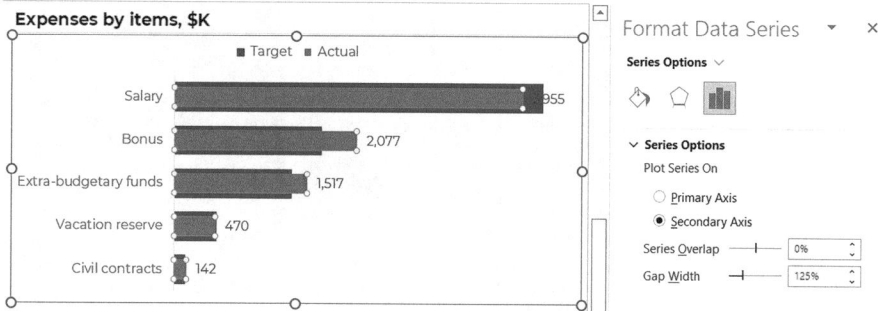

Then select the data series "Target." We change its color to neutral gray so that the emphasis remains only on the actual, and set the following settings for it: the Series Overlap is 0%, and the Gap Width is 25%.

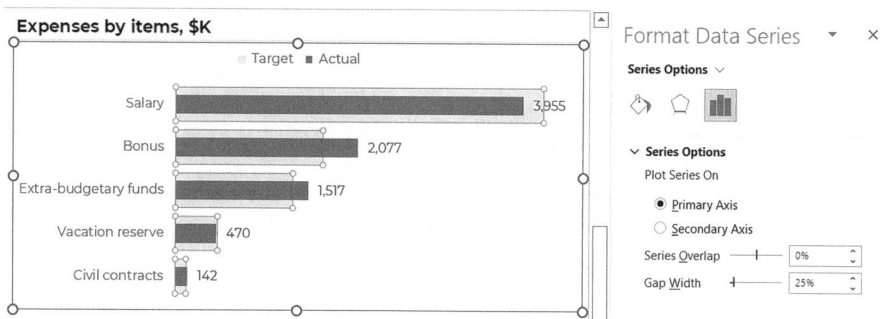

All settings of the bullet chart are completed, and it looks almost like a standard from Power BI. Similarly, you can set up a vertical bullet chart, if you need it for work.

Vertical Bullet Chart with a Dash

The second type of the standard bullet chart can also be built in Excel.

Step 1. Change the chart type by selecting the combined one, and then select the "Line with Markers" type for the target.

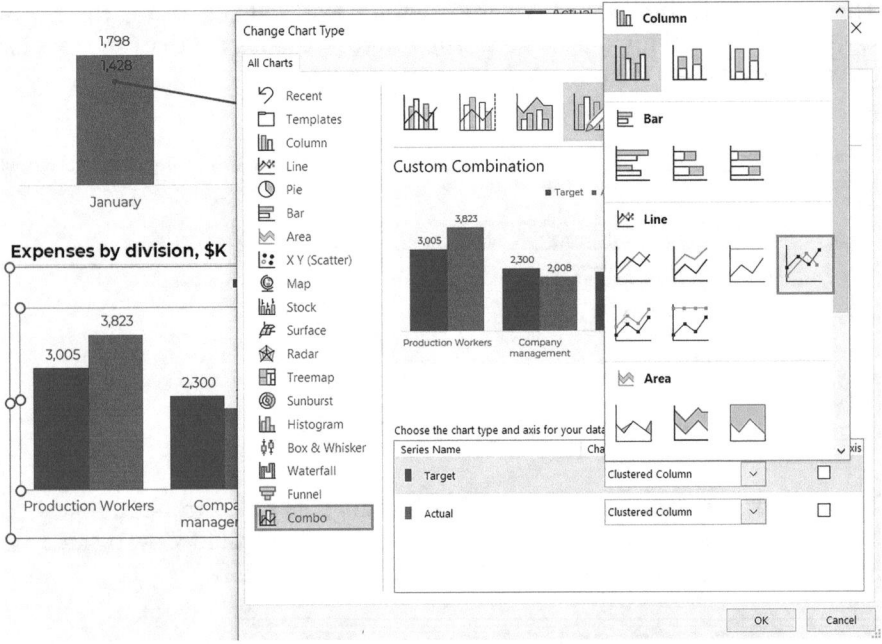

Step 2. Set up a data series "Target": in the "Line" section, set the value "No line," and in the "Marker" section in the "Marker Options" subsection, select the "Built-in" type, the type is the largest dash and set the size 45.

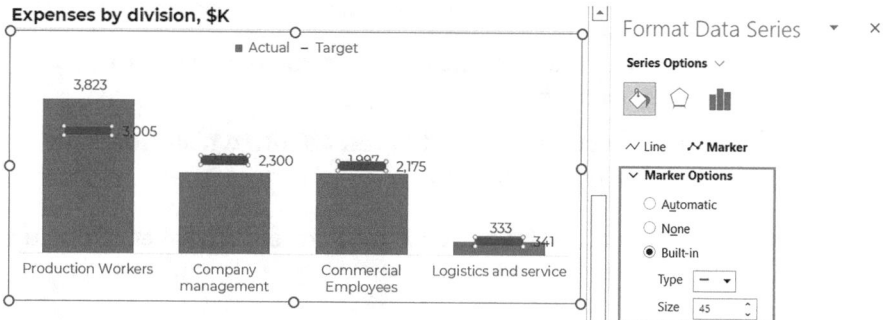

Step 3. Due to the overlap of the actual values with the marker, it is necessary to transfer the data labels for this series and select the position "In the center."

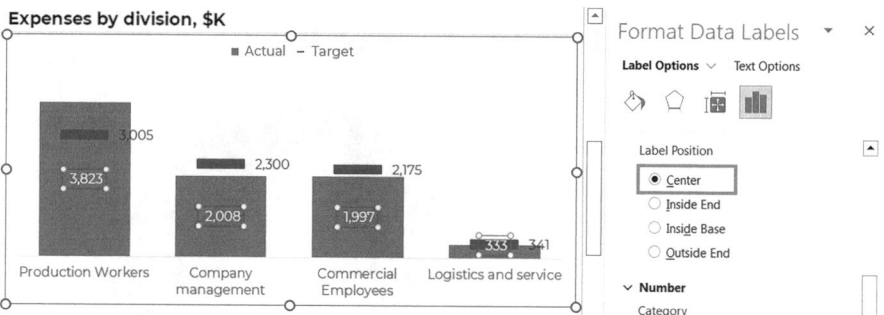

But everything turned out imperfectly: for small values, Logistics and Service still does not have enough space for data labels.

Unfortunately, this type of bullet chart can be built in Excel only for vertical, but not for horizontal columns. Because a marker is an attribute of a graph, and a graph can only display dynamics. And the dynamics can only be on a horizontal timeline. Here, oddly enough, the creators of Excel did the right thing and did not even give the opportunity to build vertical graphs.

Diagrams for a Complex Structure

In corporate reporting, the structure rarely takes any complex forms. But when this happens, about eight out of ten reports convey this information both incomprehensibly and incorrectly. In particular, this applies to two cases:

- When you need to show the structure by two parameters at once

- When the structure needs to be presented in combination with another type of analysis

I will show you which error I encounter most often, and offer alternative solutions to such problems.

Error: Doughnut Chart for Two Data Series

Traditional charts for the structure, pie chart and trimap, can display only one data series. Although two series can be displayed on the doughnut chart, and some are happy with such a life hack.

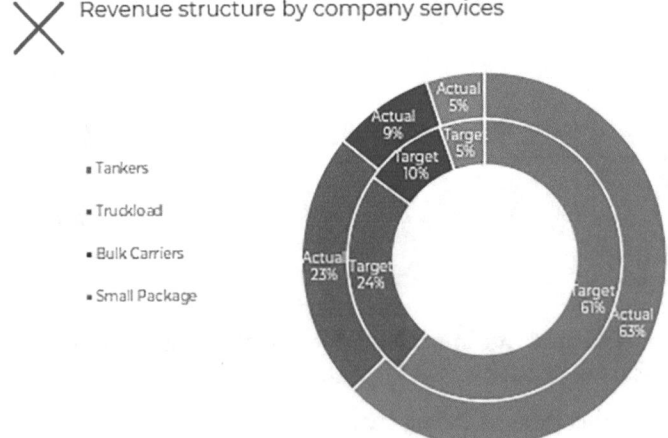

But this is still an unsuccessful visualization, it distorts the data. Look closely at the orange sector: the actual 23% looks more than the target of 24%. This can be logically explained by the fact that the radius of the ring with the actuals is larger, and therefore a smaller fraction takes up more space than on the ring with the target. But the problem is that it has to be explained. And proper visualization should work on the principle of "saw and understood."

As a compromise in such a situation, we can consider a normalized bar chart. Here two rings unfolded in a line of the same length. Although the sectors are shifted relative to each other, but 24% is clearly more than 23%. And indeed, we can see a general change in the actual revenue structure relative to the targeted one.

I draw your attention to the importance of the chart design checklist! Each item individually may seem insignificant, but in total, they create a completely different impression. Compare, if we just took and changed the type of diagram, then that would be what happened.

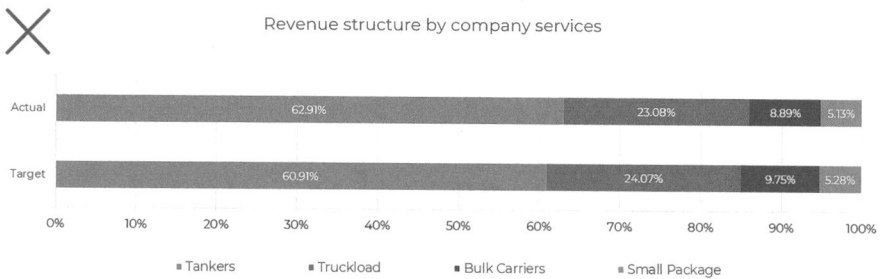

Here are the key points that we have finalized.

- Data labels were displayed as a percentage. There is no such display parameter for a bar chart, so you need to recalculate them in the source data (later you will learn how to do this in the pivot table).

- We made the stripes wide and visual by reducing the side gap, as well as by removing the scale and gridlines.

Such a diagram also has its drawbacks. For example, data labels can only be placed inside sectors here, but in small sectors they will overlap each other and become unreadable. Just like the doughnut chart, it will be visible within six sectors. However, this visualization at least helps to see the difference in the length of the columns.

Complex Trimap

Trimap can show the structure not only for one data series but also for two categories. More precisely, on two levels, for example, to group the types of services according to the activities of a logistics company. I'll tell you how to build such a diagram with fractions.

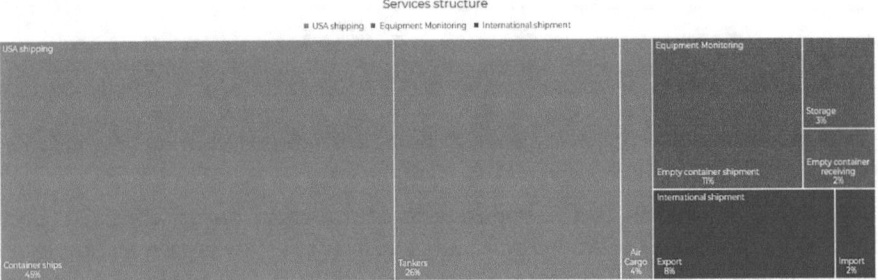

Step 1. To begin with, we will make a sample for the pivot table adding fields with areas of activity, types of services, and sales volume. By default, the pivot table is built in a hierarchical form and it is impossible to build a trimap without additional data preparation.

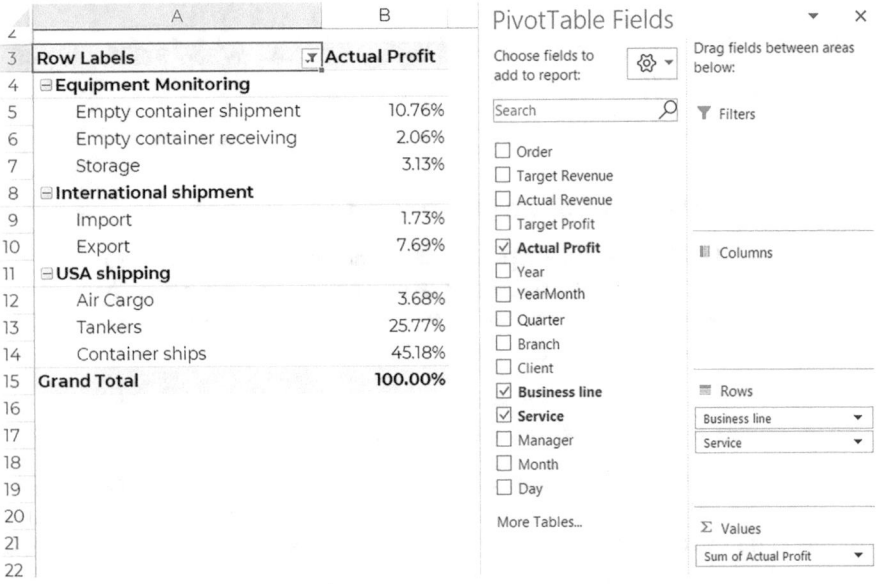

Step 2. To expand the data in the pivot table in the desired form, you need to select the "Report Layout" item on the "Design" tab to begin with, and first the "Show in Tabular Form" sub-item, and then the "Repeat All Item Labels" sub-item.

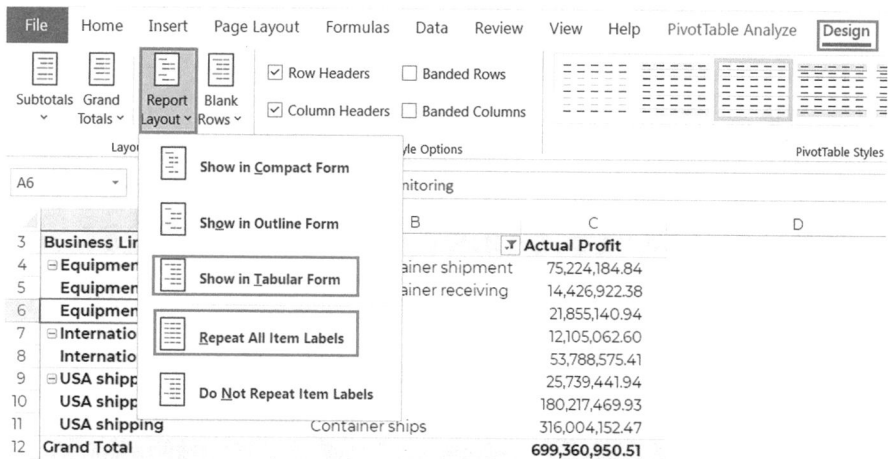

After that, disable the display of totals by clicking the "Subtotals" sub-item "Do Not Show Subtotals."

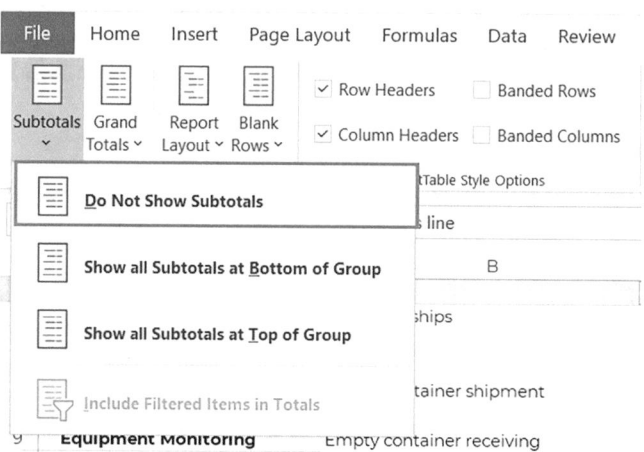

Step 3. The data is presented in absolute values and we need to convert it into percentages. To do this, go to any cell of the "Sales" column, in the right-click menu, select "Value Field Settings…." Or double-click the column header. In the window that opens, on the "Show Values As" tab, select the formula "% of Grand Total" in the drop-down menu.

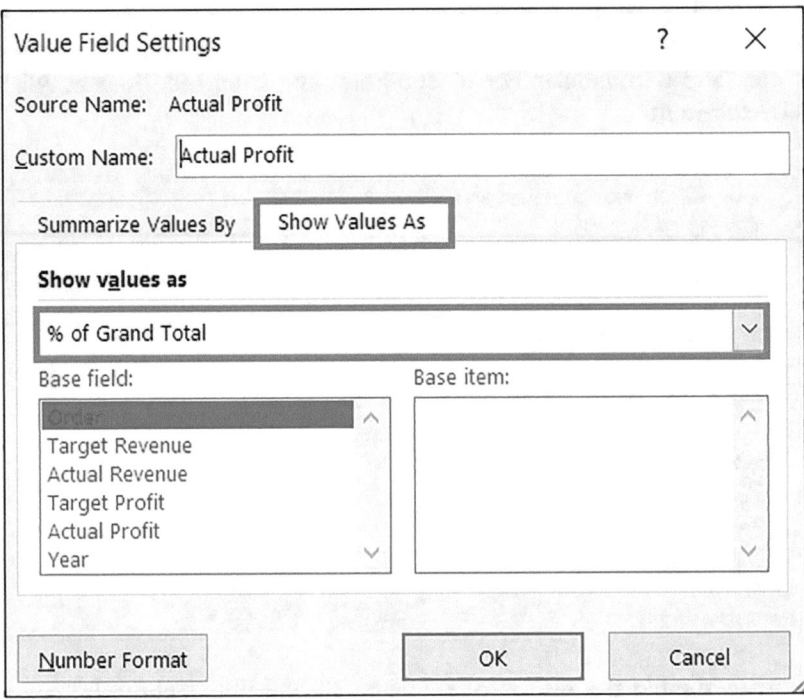

Step 4. Since it is impossible to build a treemap chart from a pivot table in Excel, we create a duplicate of its table by putting links to the cells of the pivot table. Do not forget to set the percentage data type for the column in the duplicate table (by default, the link will transmit a decimal numeric format).

	A	B	C	D	E	F	G
1							
2							
3	**Business Line**	**Service**	**Actual Profit**		Business Line	Service	
4	⊟Equipment Monitoring	Empty container shipment	10.76%		Equipment Monitoring	Empty container shipment	=C4
5	Equipment Monitoring	Empty container receiving	2.06%		Equipment Monitoring	Empty container receiving	2%
6	Equipment Monitoring	Storage	3.13%		Equipment Monitoring	Storage	3%
7	⊟International shipment	Import	1.73%		International shipment	Import	2%
8	International shipment	Export	7.69%		International shipment	Export	8%
9	⊟USA shipping	Air Cargo	3.68%		USA shipping	Air Cargo	4%
10	USA shipping	Tankers	25.77%		USA shipping	Tankers	26%
11	USA shipping	Container ships	45.18%		USA shipping	Container ships	45%
12	**Grand Total**		**100.00%**				

Step 5. We create a new treemap chart based on the duplicate table and configure the data labels by ticking the category name and value items, as well as selecting the "New Line" separator type.

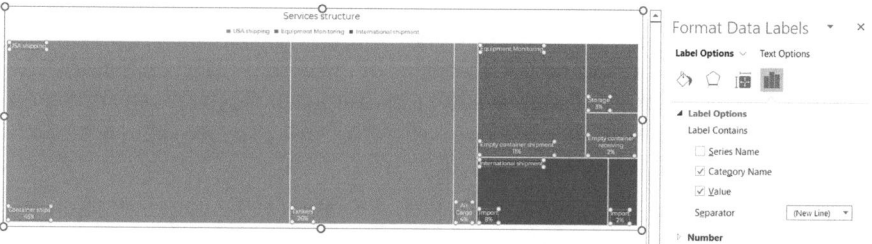

We have reviewed the techniques of how to make non-standard visualization from simple diagrams. To be honest, I think it's more like crutches than life hacks: adjust auxiliary axes, hide the scale in the background color, disable lines from the graph, and put links from the pivot table to neighboring cells.

If you often do this in Excel or PowerPoint, then maybe it's time for you to switch to Power BI. There, these visualizations have already been enhanced, and there are many more useful diagrams and functions for working with data.

On the other hand, with understanding the anatomy of diagrams at this level, you will have more freedom of action in advanced BI systems.

Summary

Usually, an actual is compared with a target or other indicator on a dashboard. To keep the visualization clear, we use non-standard diagrams. In modern BI systems, these are ready-made widgets, but in Excel you have to use tricks to get a similar result.

Rating: We combine two columns according to the bullet chart principle, as if the actual "breaks through" the boundary of the target.

- Based on a bar chart – by constructing a series of actuals along the auxiliary axis, adjusting the gap width.

- Based on a column chart – by displaying a series of the target in the form of a linear graph with markers. And then we remove the line and adjust the marker in the form of a large strip.

Dynamics

- Combine a line and a column to avoid piling up points in one place.

- Use the auxiliary axis if the values are of different order (one in hundreds and the other in thousands). Only the scale cannot be deleted, you will have to mask it in the background color.

Structure

- To compare the structure of the target and the actual, we use a 100% stacked bar chart. It will be visual within six categories.

- The Treemap chart is suitable for two categories. It will show the overall structure by top-level categories, as well as the structure of subcategories within.

6.4 How to Show Everything at Once

Data visualization projects usually start with a nice neat layout. But the result most often looks quite different. Once the customer sees the product, new requirements and ideas appear.

– And let's add to the diagram not only the target but also the actual value of last year?

– It is not enough to compare with last year (a pandemic, a crisis or a new disaster). We also need the previous year.

– Add percentage deviations to the graph. No, in absolute values. Hmm, and let's have both in percentages and in absolute values!

This list can be continued for a long time, and every new item makes managers and analysts' eyes twitch. Of course, it turns out to be terrible.

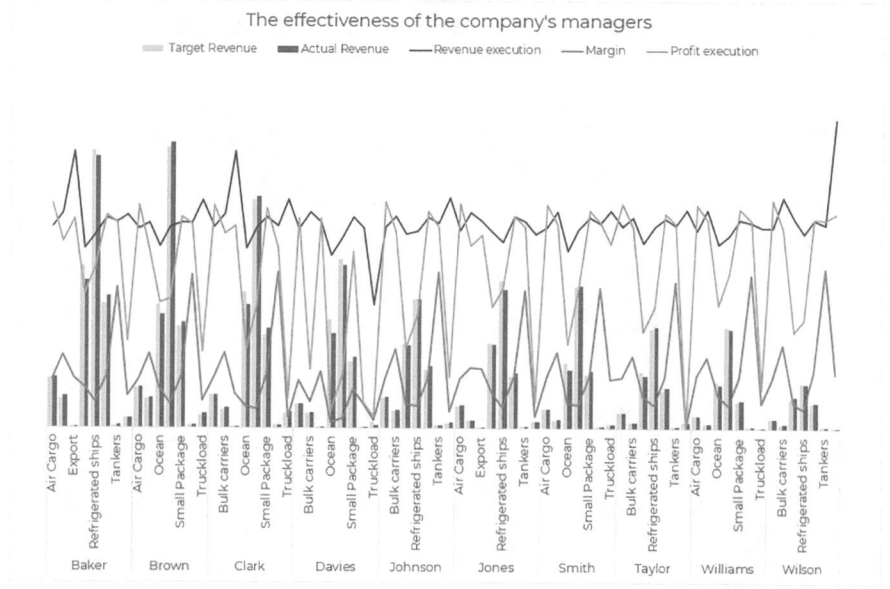

Sometimes such a problem arises because we forget what kind of analysis we use: somewhere we change the type of diagram, somewhere we divide it into two. But sometimes you really need to compare several indicators at once.

Of course, the last word always remains with the customer. In my company's projects, sometimes you also have to compromise, which I will not show in the portfolio as a best practices example.

When a Table Is the Only Option

Even in Power BI and Tableau, there is no chart that will visually display four rows of data in the context of several categories. Three indicators is the limit for a single chart: you can use the combined charts that I showed in the previous part, or a two-level treemap chart for categories with subcategories.

For a larger amount of data, there is only one working option – tables. Many people don't like this option, but at least it will allow you to read the values. And it's better than an overloaded diagram.

■ **Tip!** *To make the table visual and easy to understand, it is important not to overdo it with colorful design and to place visual accents correctly.*

Many are content with the standard formatting of pivot tables. But my approach here is the same as with the diagram: there is a draft for calculations, and there is a final version that is on a sheet with a dashboard.

Manager	Actual Revenue	Target Actual	Actual Profit	Target Profit	Margin	Number of orders
Baker	109,210,465	110,674,500	21,750,454	25,324,600	20%	4,616
Brown	99,961,003	99,554,850	18,889,262	24,565,900	19%	3,982
Clark	89,147,270	88,851,450	14,092,176	20,255,500	16%	3,519
Davies	63,206,326	65,678,250	5,748,155	16,182,100	9%	2,457
Johnson	56,067,386	55,170,500	9,912,982	14,127,700	18%	2,367
Jones	53,815,513	54,551,500	9,244,216	12,076,400	17%	2,145
Smith	49,714,003	49,715,750	8,562,733	11,145,200	17%	2,036
Taylor	38,044,022	37,594,000	6,603,661	9,652,400	17%	1,429
Williams	32,653,667	33,167,250	5,723,897	6,870,600	18%	1,297
Wilson	20,596,789	19,852,750	3,819,020	5,001,500	19%	884
Grand Total	**612,416,444**	**614,810,800**	**104,346,557**	**145,201,900**	**17%**	**24,732**

Manager	Revenue, $M	Revenue, exec.	Profit, $M	Profit, exec.	Маржа	Number of orders
Baker	109	99%	22	86%	20%	4,616
Brown	100	100%	19	77%	19%	3,982
Clark	89	100%	14	70%	16%	3,519
Davies	63	96%	6	36%	9%	2,457
Johnson	56	102%	10	70%	18%	2,367
Jones	54	99%	9	77%	17%	2,145
Smith	50	100%	9	77%	17%	2,036
Taylor	38	101%	7	68%	17%	1,429
Williams	33	98%	6	83%	18%	1,297
Wilson	21	104%	4	76%	19%	884
Grand Total	**612**	**100%**	**104**	**72%**	**17%**	**24,732**

■ **Note** In the designed version, I have removed the targeted indicators. And instead of them, I added percentage deviations with color coding according to the principle of traffic lights. There is no need to calculate the difference between the target and the actual in your mind – we have a visual emphasis on the target and a visual indicator of deviation from the target.

Let's look at the steps of how to get such a result using the additional options of pivot tables.

How to Add a Calculated Field

In my example, it will be calculations for the target execution for revenue, profit, and margin.

We add columns to the pivot table on the tab "PivotTable Analyse": section "Calculations" ➤ button "Fields, Items & Sets" ➤ item "Calculated Field…."

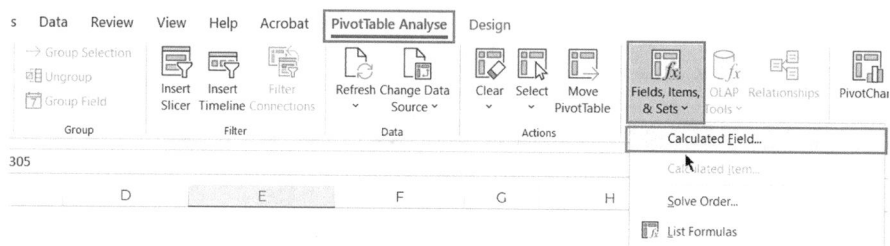

In the "Insert Calculated Field" window that opens, we give a clear name to the new field and set the formula for calculations. If the formula needs data from existing fields of the pivot table, select them from the "Fields" block by name.

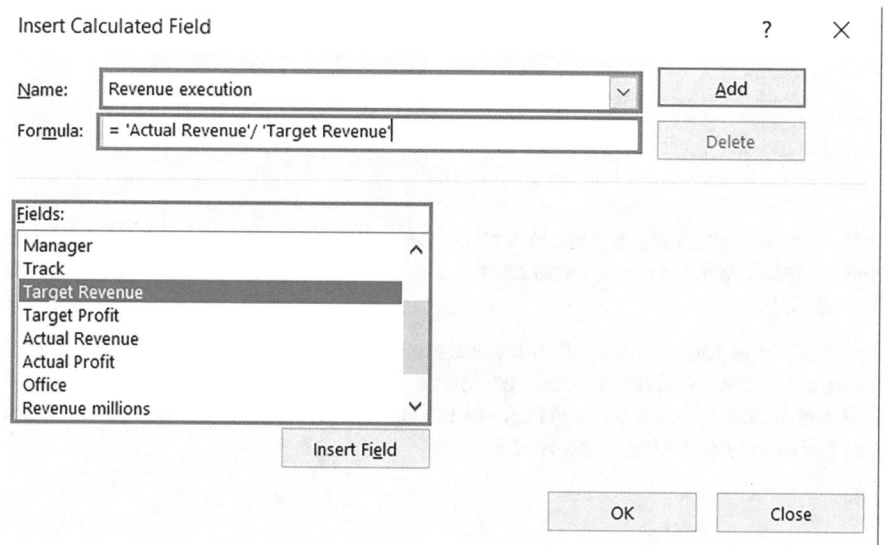

You also need to change the bit depth of the actual indicators by millions – so the numbers will become shorter and it will be easier to perceive them. To do this, we create calculated columns in the same way, only in the formula we divide the value from the "Actual" column by 1,000,000.

As a result, three new indicators appeared in the fields list (revenue, profit, and margin), as well as two fields with revenue and profit converted into millions. Someone will say that it is easier and faster to divide a fact into a plan in a neighboring cell, and he will be right. But the advantage of the calculated field is that it can be used in other selections of the pivot table. This approach will save time in the future, and in general it is considered good form.

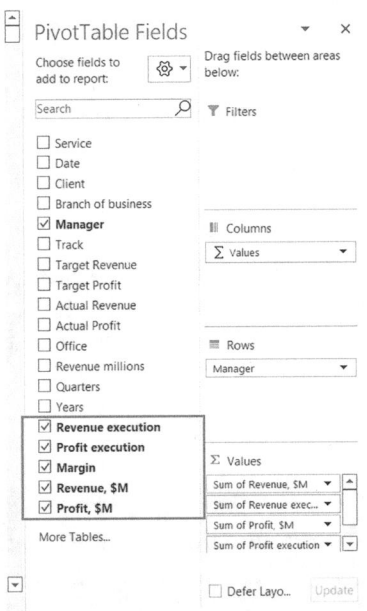

We removed the target from the table and added calculated fields. But it's still just a table with no visual accents. To arrange them, we use conditional formatting.

In Excel, this functionality is very extensive and has a bunch of customized formats. But we will not choose its formatting styles – there will be a primitive like red filling of cells with green values or other visual ugliness. We will set the necessary settings ourselves.

Rules for Formatting Cells

Before formatting the table, we will determine its rules in advance. To do this, we need to determine the formatting methods (where to add histograms, where to add icons, and where to change the font color), as well as set the values at which the color will change.

1. For clarity, we will add histograms to the numerical indicators of revenue, profit, and the number of orders – the length of the columns in the cells will make the comparison visual.

2. Columns with percentages are formatted using the font color. For the column with the target execution, we use the following rule: if the target is completed, the color is green, if not, red.

3. The values in the "Margin" column are marked with icons with traffic light windows. Negative indicators will be highlighted in red, average ones in yellow, and positive ones in green.

Conditional Formatting Using Histograms

For columns with actual values, histograms are usually added to visually show the rating.

Select the column and, on the "Home" tab, under the "Conditional formatting" button, select the "Histograms" item, and then specify its type. For the column "Revenue, $M.", we will add a blue histogram.

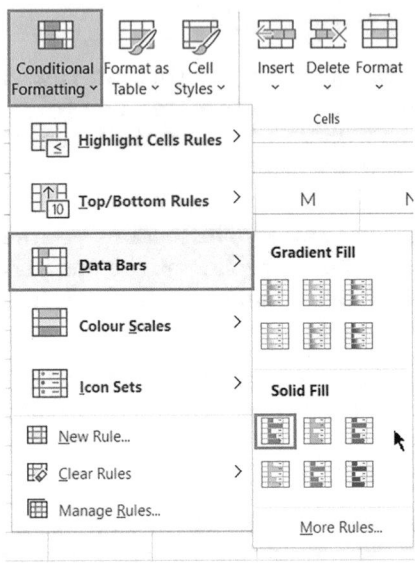

The color of the added histogram is saturated by default, but it is better to use less bright shades for formatting tables. They can be changed after setting up all the conditions.

To do this, on the "Home" tab, under the "Conditional Formatting" button, select the "Manage Rules…" item. The Conditional Formatting Rules Manager window that opens already shows the rules that are applied to the pivot table.

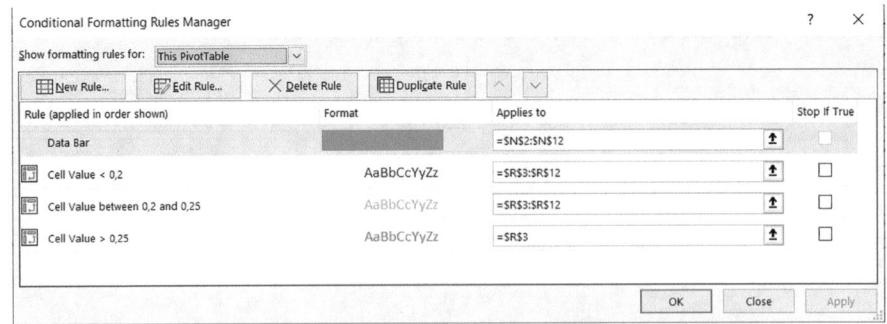

We choose the necessary one from the rules: you can just double-click it or click "Edit Rule…." In the window that opens, in the "Color" section, set a lighter shade of blue by moving the slider up. Click "OK."

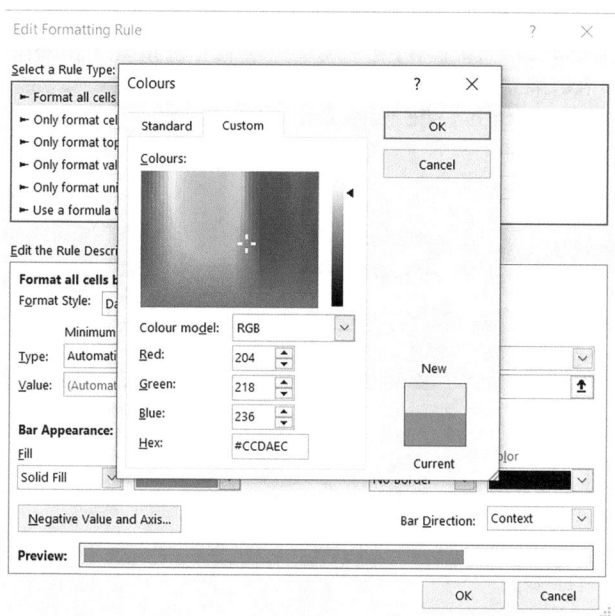

Similarly, we change the color for other histograms. Select the "Revenue execution" column and on the main menu tab, under the "Conditional Formatting" button, select " Highlight Cells Rules" ➤ "Greater Than…."

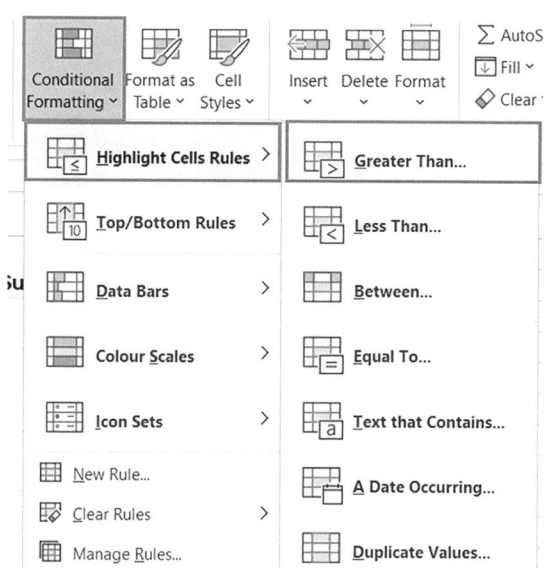

In the window that opens, specify the value in the first field of the section "Format cells that are GREATER THAN" – 100%. In the second field, select "Customised format." In the additional window "Format Cells" that opens, we set the green color on the Font tab.

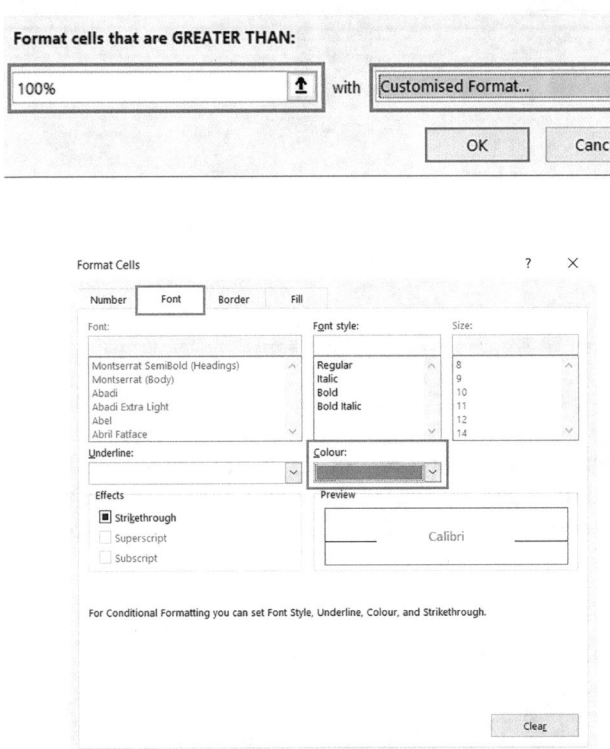

Now we will set the second rule for the same column – for cases when the target is not achieved. Repeat the previous steps, only in the menu "Conditional formatting," select the item "Less Than…," and in the "Customised Format…" set the red color.

Conditional Formatting with Icons

For conditional formatting of the "Margin" column, we will use colored icons according to the traffic light rule:

1. Red color – for values less than 10%

2. Yellow color – for values from 10% to 20%

3. Green color – for values greater than 20%

To do this, select the "Margin" column and select "Icon Sets" ➤ "More Rules…" on the main menu tab under the "Conditional Formatting" button.

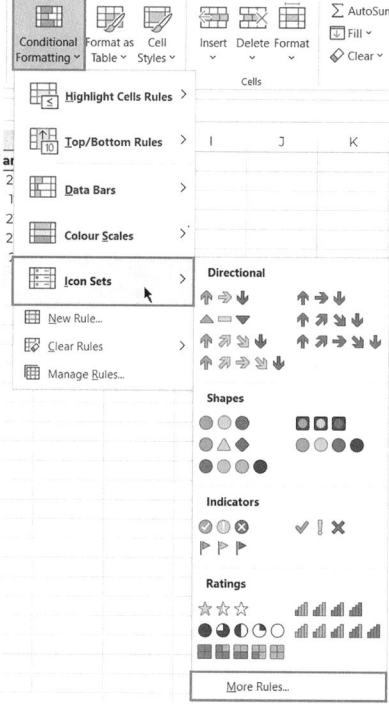

In the window that opens, we set our rules. Specify the desired value in the "Display each icon according to these rules" section, and select "Number" in the "Type" drop-down list.

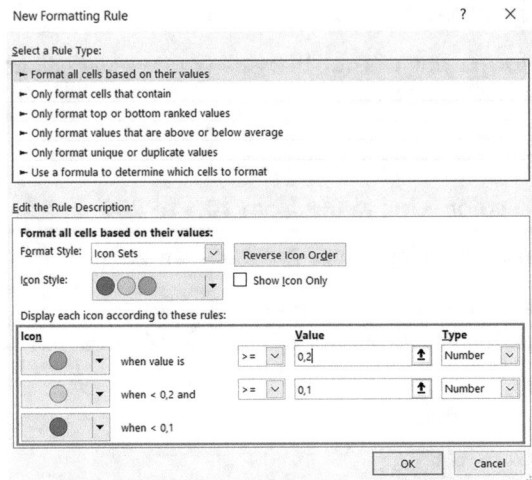

After all these actions, we will remove the fill for the headers and totals. And we will get a harmoniously designed table in calmer colors.

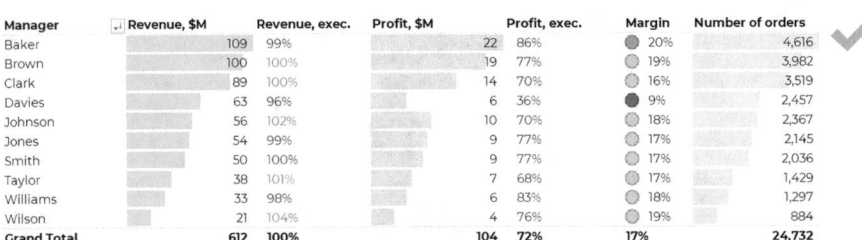

Since this is a pivot table, you can connect slicers to it to filter data.

Typical Errors

With conditional formatting, it is important not to overdo it, otherwise it will start to ripple in the eyes. For example, this happens when both the value and the background of the cell are colored. You need to know the limit, and it's like this: **one column is one way of formatting.**

Manager	Revenue execution
Williams	98%
Wilson	104%
Smith	100%
Clark	100%
Baker	99%
Jones	99%
Brown	100%
Davies	96%
Johnson	102%
Taylor	101%
Total	**100%**

Manager	Revenue, exec.
Williams	98%
Wilson	104%
Smith	100%
Clark	100%
Baker	99%
Jones	99%
Brown	100%
Davies	96%
Johnson	102%
Taylor	101%
Total	**100%**

Manager	Revenue actual	
Baker	⇑	109,210,465
Brown	⇑	99,961,003
Clark	⬈	89,147,270
Davies	⇒	63,206,326
Johnson	⇒	56,067,386
Jones	⬊	53,815,513
Smith	⬊	49,714,003
Taylor	⬇	38,044,022
Williams	⬇	32,653,667
Wilson	⬇	20,596,789
Total		**612,416,444**

Manager	Revenue, $M
Baker	109
Brown	100
Clark	89
Davies	63
Johnson	56
Jones	54
Smith	50
Taylor	38
Williams	33
Wilson	21
Total	**612**

To choose the correct formatting method, think about the types of data analysis. If you need to compare the values and show what is more and what is less, this is a rating. It is displayed in the form of horizontal columns. And deviations in percentages are highlighted according to the principle of a traffic light.

A separate love of novice designers is a gradient. I use it extremely rarely for two reasons.

- **For logical reasons.** We set specific threshold values for indicators: "good" – in the green zone, "bad" – in the red. We assign yellow color to intermediate permissible deviations, and do not smear them in a gradient.

- **For aesthetic reasons.** The gradient between red and green gives a dirty shade. And together with yellow, it becomes either dirty or too pale.

Manager		Margin
Baker	●	20%
Brown	●	19%
Wilson	●	19%
Johnson	●	18%
Williams	●	18%
Taylor	●	17%
Smith	●	17%
Jones	●	17%
Clark	○	16%
Davies	●	9%
Total		**17%**

Manager		Margin
Williams	●	18%
Wilson	●	19%
Smith	●	17%
Clark	●	16%
Baker	●	20%
Jones	●	17%
Brown	●	19%
Davies	●	9%
Johnson	●	18%
Taylor	●	17%
Total		**17%**

An exception for gradient fill can be the "heat map" technique. The following example shows how managers are busy working with clients by month. The situation is ambiguous: on the one hand, it is clear who has a lot of work and who has little; on the other hand, the shades begin to ripple in the eyes.

Workload of managers by month

Manager	Jan	Feb	Mar	Apr	May	Jun	Jul	Aug	Sep	Oct	Nov	Dec	Total
Baker	386	359	383	381	603	384	401	390	406	392	391	380	4,856
Brown	345	301	324	334	359	347	337	321	350	318	313	333	3,982
Clark	288	299	290	304	309	314	286	283	306	271	303	266	3,519
Davies	223	222	209	305	207	215	204	80	212	222	280	213	2,592
Johnson	201	381	195	216	195	200	189	197	199	185	208	189	2,555
Jones	183	163	162	181	325	179	204	215	158	54	173	178	2,175
Smith	172	175	162	163	168	162	196	177	181	169	146	165	2,036
Taylor	112	127	122	174	115	126	135	138	117	105	129	118	1,518
Williams	115	68	91	102	202	200	173	115	121	99	108	117	1,511
Wilson	63	88	89	75	70	69	70	71	78	55	86	70	884

In such a situation, it is worth considering which zones to focus on: negative or positive. There is no need to go to the customer with questions about the color design: it is better to prepare and show possible options.

Option 1. Positive

The most saturated fill is used for the maximum value – it is immediately clear that the peak of sales was in May at the manager Baker. Similarly, peaks or literally voids in sales are noticeable. And it is obvious whose contribution to the overall result is invisible.

Workload of managers by month

Manager	Jan	Feb	Mar	Apr	May	Jun	Jul	Aug	Sep	Oct	Nov	Dec	Total
Baker	386	359	383	381	603	384	401	390	406	392	391	380	4,856
Brown	345	301	324	334	359	347	337	321	350	318	313	333	3,982
Clark	288	299	290	304	309	314	286	283	306	271	303	266	3,519
Davies	223	222	209	305	207	215	204	80	212	222	280	213	2,592
Johnson	201	381	195	216	195	200	189	197	199	185	208	189	2,555
Jones	183	163	162	181	325	179	204	215	158	54	173	178	2,175
Smith	172	175	162	163	168	162	196	177	181	169	146	165	2,036
Taylor	112	127	122	174	115	126	135	138	117	105	129	118	1,518
Williams	115	68	91	102	202	200	173	115	121	99	108	117	1,511
Wilson	63	88	89	75	70	69	70	71	78	55	86	70	884

Option 2. Negative

The situation is reversed here: the brightest red color is at the minimum value, and the maximum is highlighted in white. But with this approach, almost the entire table is lit red, even the leaders in terms of sales. Not the most successful visualization, because the accents were lost again.

Workload of managers by month

Manager	Jan	Feb	Mar	Apr	May	Jun	Jul	Aug	Sep	Oct	Nov	Dec	Total
Baker	386	359	383	381	603	384	401	390	406	392	391	380	4,856
Brown	345	301	324	334	359	347	337	321	350	318	313	333	3,982
Clark	288	299	290	304	309	314	286	283	306	271	303	266	3,519
Davies	223	222	209	305	207	215	204	80	212	222	280	213	2,592
Johnson	201	381	195	216	195	200	189	197	199	185	208	189	2,555
Jones	183	163	162	181	325	179	204	215	158	54	173	178	2,175
Smith	172	175	162	163	168	162	196	177	181	169	146	165	2,036
Taylor	112	127	122	174	115	126	135	138	117	105	129	118	1,518
Williams	115	68	91	102	202	200	173	115	121	99	108	117	1,511
Wilson	63	88	89	75	70	69	70	71	78	55	86	70	884

Option 3. Interactive

Analysts often place all sorts of table combinations on several sheets. And they call it a dashboard, although, with such a presentation, its idea is lost: you need a separate instruction on where to look and where to start.

Even if the customer insists on such a table, I offer him an option in the ideology of an interactive dashboard. And I definitely show him how to use it.

For example, the chart shows that in August we had a peak in sales. Whose merit is this? We choose August in a slicer and see that the leader in revenue this month is manager Williams, and Jones brought the least money to the company.

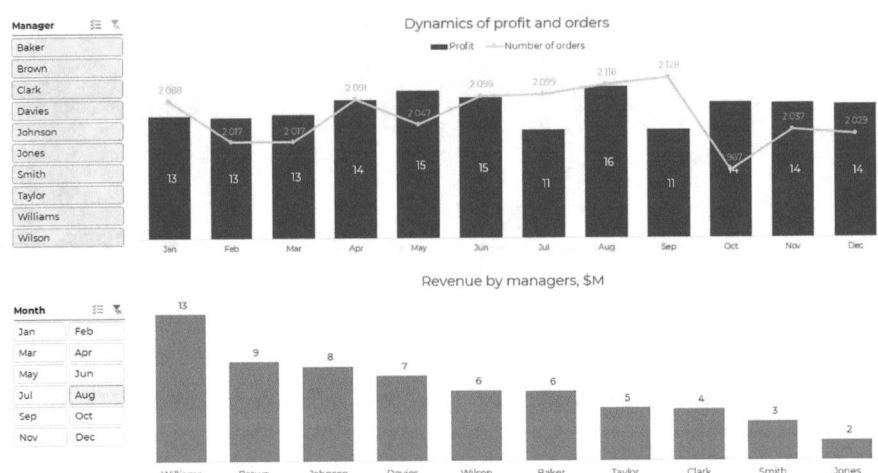

We look at Williams' personal statistics by selecting his surname on the slicer. We see that he is the best for the year. During the year, manager Williams brought 125 million dollars in revenue, and the company's profit from his work averaged two million dollars per month.

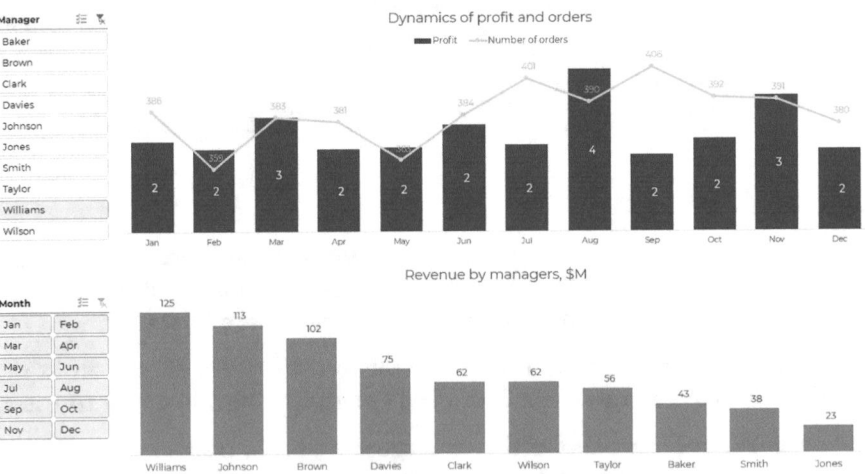

I often come across managers who don't really need visualization and tables are quite enough. This is especially true for people with a financial background: one table will tell them more than a dozen charts. This group of managers will especially appreciate the use of these rules, because they save a lot of time on independent data analysis.

Summary

Sooner or later, any analyst has to display such a number of indicators on the chart that it becomes unreadable. A good solution for such cases is conditional formatting of table fields. It exists in all BI-systems and the principle of operation in them is the same as in Excel.

Five rules for placing visual accents

1. Instead of target, actual, and deviation indicators, use only actual and deviation. Take away what doesn't tell you anything new.

2. Design the actual in the form of colored bars, and for deviations use green, yellow, and red colors. It can be a font color or an icon with a traffic light circle.

3. If you work with pivot tables, add calculated fields to calculate deviations. They can also be used in other data samples.

4. Use only one formatting technique per column: data bars, font color, traffic light icons, or cell fill.

5. Make sure that the formatting highlights the peak values, and not just paints the table in all the colors of the rainbow.

6.5 Funnel, Waterfall, and Magic Pillow

In Part 6.3, I describe life hacks for displaying multiple data series on charts, as well as how to build a non-standard bullet chart based on regular columnar charts. But there are also advanced charts that can be built in Excel.

The most popular of the non-standard ones are the funnel and waterfall diagrams. Salespeople and marketers dream of visualizing a funnel, and financiers dream of factor analysis on a waterfall diagram.

For a long time, many people believed that it was impossible to build them in Excel. Someone used additional add-ins, for example, Think-cell, to make such a diagram in PowerPoint. Someone thought that only a BI system was needed for this, otherwise there was no way. And there are two nuances here:

- Both the funnel and the waterfall are in the new versions of Excel, but it is impossible to build them on pivot tables.

- Both diagrams can be made in older versions with the help of "pillows" – auxiliary transparent rows.

Funnel = Structure + Dynamics

Initially, the sales process was visualized using a funnel, divided into stages. There are a lot of calls and cold contacts at the entrance, some of them go into meetings, presentations, and so on before signing the contract.

On the one hand, it is a sequential process, that is, dynamics. And on the other – the analysis of changes in the structure: we look at what proportion is eliminated at each stage. A Funnel diagram is suitable for combining these two types of data analysis.

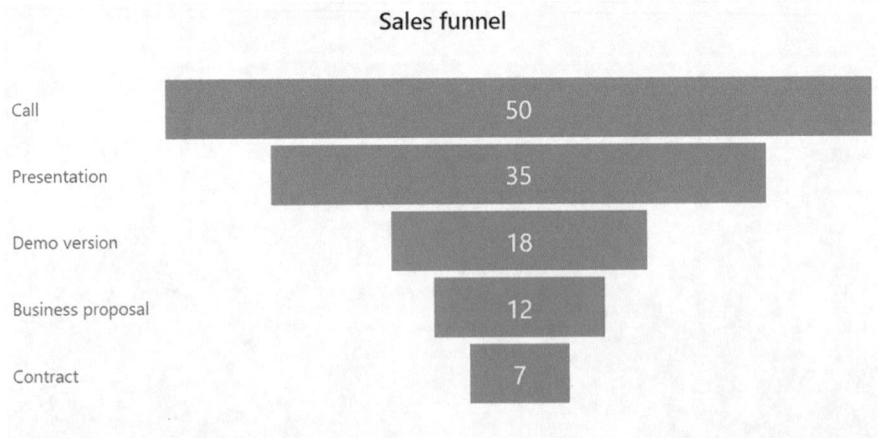

Sales funnel

Call	50
Presentation	35
Demo version	18
Business proposal	12
Contract	7

In HR, the same approach is used to visualize the recruitment process. It is important as well that the funnel is constantly replenished with new candidates. And it is also known that at each subsequent stage, there will be a dropout. Here is an example of a diagram built in Power BI – I like it that you can immediately display the conversion of each stage.

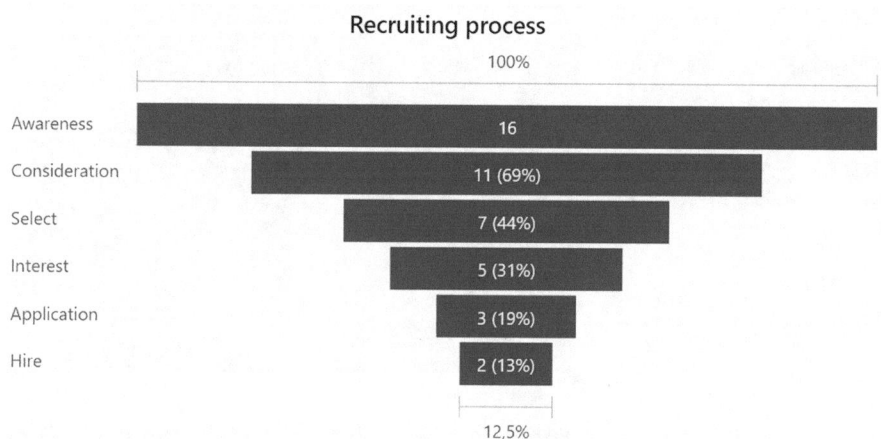

How to Build a Funnel in Excel 2019 and Above

Starting from version 2019, this chart is already available in the standard Excel set. I was pleasantly surprised by its default view: minimal side clearance, contrasting labels. Everything is fine, but Excel refuses to build it on pivot tables. But this chart can be made interactive with the possibility of filtering by slicer.

Stage		Stage	Funnel
Call	40	Call	40
Presentation	14	Presentation	14
Demo version	10	Demo version	10
Proposal	8	Proposal	8
Contract	5	Contract	5

Manager ⅀≡ ⦿		Sales process	
Alex		Call	40
Mary		Presentation	14
Pavel		Demo version	10
Victor		Proposal	8
		Contract	5

It turns out that the slicer filters the pivot table on the left, and the cells on the right refer to it. So we outwitted Excel and built a clickable funnel.

Life Hack for Older Versions

Not everyone has Excel 2019 or Office 365, so I'll show you how to build such a chart in any old version. This can be done using an additional column in the table and a bar chart.

In fact, the funnel is a bar chart, only aligned in the center. Excel does not allow you to adjust such alignment, and this is exactly what the auxiliary calculation column – the so-called "pillow" will help. In it, we will indicate how much the data column needs to be moved away from the category axis so that the diagram takes the form of a funnel. In the future, we will make this pillow invisible.

Step 1. We add a new column with the name "Pillow" to the table. In each row of this column, you need to set values using a formula. In the first row with the largest number, this value is zero, because this column of the chart does not need to be moved anywhere.

For each subsequent stage, in the "Pillow" column, we set the formula: divide the difference in the values of the funnel at the previous and current stages by 2 and add the value of the pillow from the previous stage.

=(E2-E3)/2 +F2

D	E	F
Stage	**Funnel**	**Pillow**
Call	40	
Presentation	14	=(E2-E3)/2 +F2
Demo version	10	15
Proposal	8	16
Contract	5	17,5

In our case, the formula looks like this: **=(E2-E3)/2+F2**, where E is a column with numbers, F is a column with pillow values, and 2 and 3 are row numbers with the necessary data.

You cannot add the previous value every time, but always subtract from the first, the largest (I have this number of calls). To do this, to maintain cell E2 (we put a $ sign in front of the letter and digit or just press F4).

=($E$2-E3)/2 +F2

D	E	F	G
Stage	**Funnel**	**Pillow**	
Call	40		
Presentation	14	=($E$2-E3)/2 +F2	
Demo version	10	15	
Proposal	8	16	
Contract	5	17,5	

Step 2. Now there are three columns in our table. We select them and insert a stacked bar chart from the menu ribbon. What Excel will build at this stage does not look like a funnel at all yet.

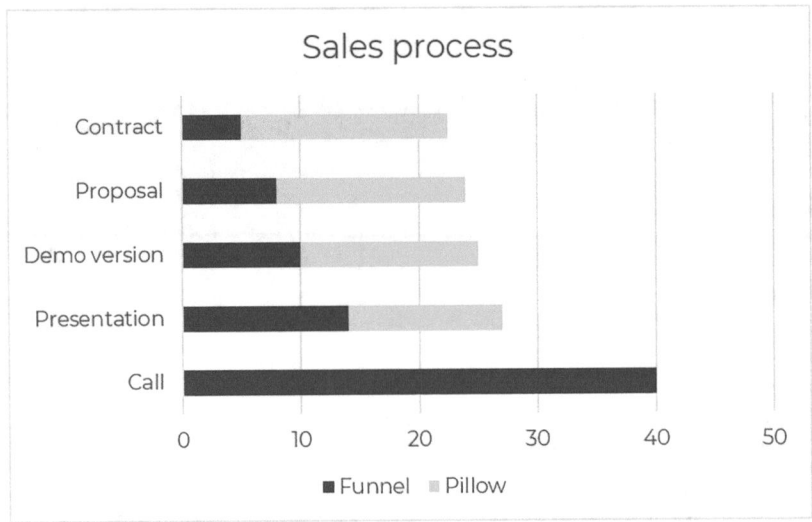

The fact is that the pillow should go first and move the funnel to the center. To do just that, we call the "Select Data…" menu and in the "Select Data Source" window that opens, we change the order of the columns: we lower the "Funnel" row down by clicking the arrow button.

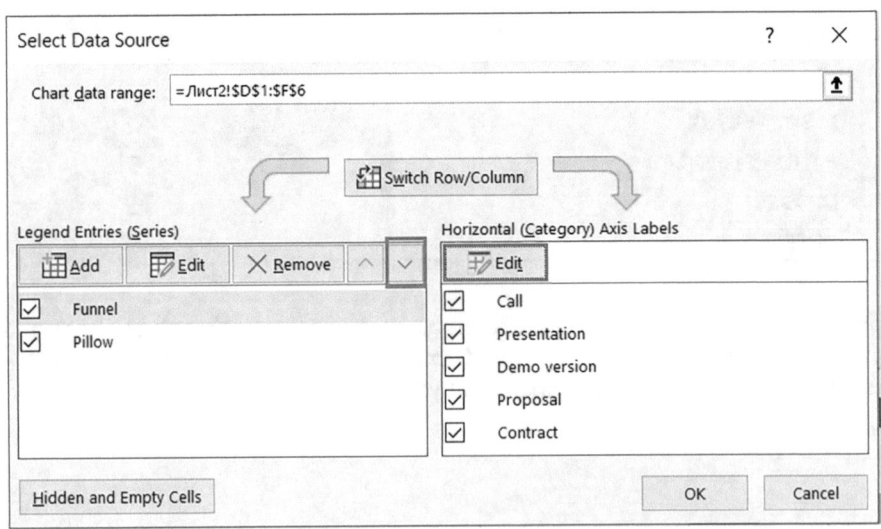

The second step, I hope you have already learned. It is necessary to check the "Categories in reverse order" on the axis for the hundredth time. Then it will become more like a prototype of the desired diagram.

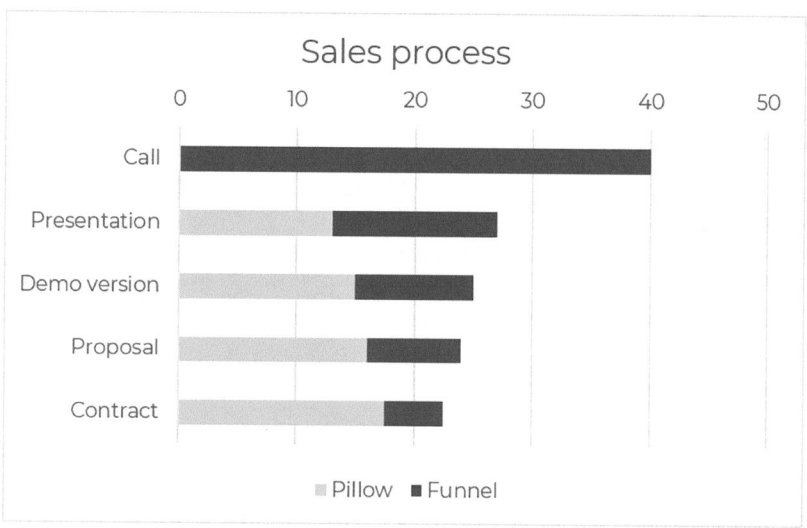

Step 3. In order for the diagram to take the shape of a funnel, now we need to make the "Pillow" row invisible, that is, simply remove the color fill. All that remains is to set up the funnel according to the checklist from Part 3.4:

- Adding data labels
- Removing the axis with the values
- Removing the gridlines
- Setting a gap width of 10% for the data series
- Removing the legend
- Increasing the font size of inscriptions and data labels to 12

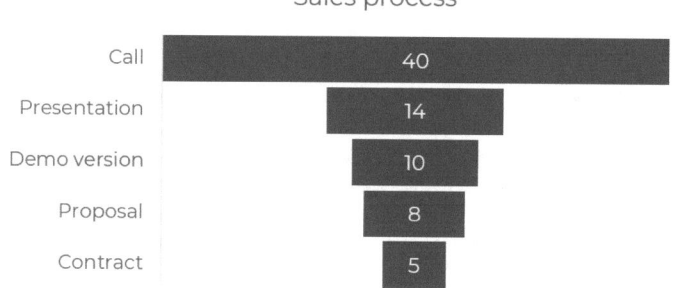

Waterfall = Dynamics + Rating

The waterfall chart is also known as a flying breaks chart or Mario kart due to the apparent suspension of columns (bricks) in midair. Often in finance, it will be referred to as a bridge.

Most often, this visual element is used for factor analysis. The diagram consists of columns: the first and last show the initial and final values, and the intermediate ones show the factors that influenced the result. Clarity is given by the different colors of the intermediate columns: green – where the indicator has grown; red – where it has decreased.

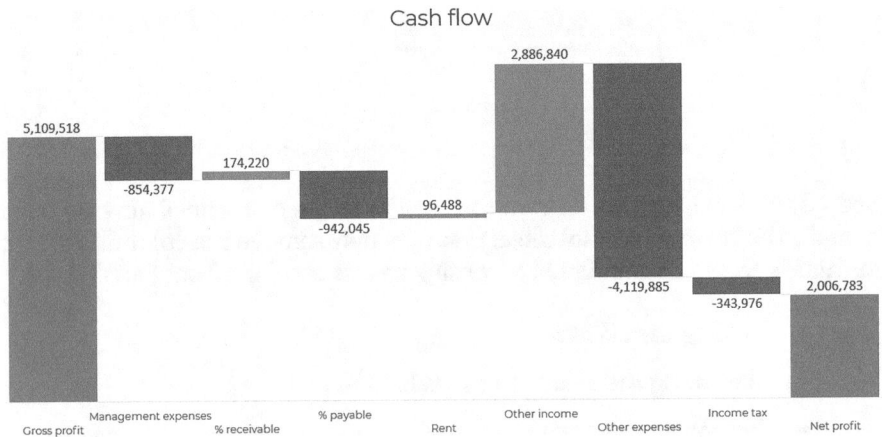

This example is based on financial data about how the net profit was formed. We see that management expenses were deducted from the gross profit, interest was added to the receipt, and interest was deducted for payment. Then they added 2.8 million other income and received a peak cash flow: 4.1 million of it was written off to other expenses, and 340 thousand taxes were paid from the remaining small base. After that, the resulting gray column of two million net profit remained.

To get such a visualization, we also combine two basic types of analysis:

- Dynamics in order to display the changes in the indicator from the starting point to the end point by stages
- Rating to quantify the factors that influenced the result positively or negatively

The financier will understand this information in the table. But for a manager, an entrepreneur who is not immersed in the details of calculations, such visualization allows one to immediately see why the company has losses, and what influenced the increase in profits.

The color coding can be inverted in the case of cost analysis. Here we see that the planned labor costs were 55 million, and in fact amounted to 70.5 million. The main savings came out due to the shift of deadlines, the withdrawal of facilities, and a little due to the reduction of administrative and managerial staff. And the overspending came out due to the revision of rates, turnover, and bonuses. We brought the most significant factors to the diagram, but there were also small ones that saved five million dollars in total.

I will tell you how to build a diagram in newer versions of the program (spoiler is easy), as well as how to use a life hack with a "pillow" in versions before 2016.

How to Build a Waterfall Chart in Excel 2016 and Above

In Excel, this chart has appeared since 2016, but it is rarely used. Largely because of the construction logic: it is not intuitive, and it cannot be formatted according to the pattern of the columns and graphs we are familiar with.

Let's look at the financial example about the profit change. The initial data should be written to the table so that the expenses go with a minus sign. It is by this criterion that Excel will determine the direction of the column: positive values – up, negative values – down.

Item	Sum
Gross profit	5,109,518
Management expenses	- 854,377
% receivable	174,220
% payable	- 942,045
Rent	96,488
Other income	2,886,840
Other expenses	- 4,119,885
Income tax	- 343,976
Net profit	2,006,783

Step 1. On the ribbon we find an icon with a waterfall chart. Click and get a strange shape. At this step, many are finally disappointed in the new Excel and go to collect a waterfall of rectangles in PowerPoint.

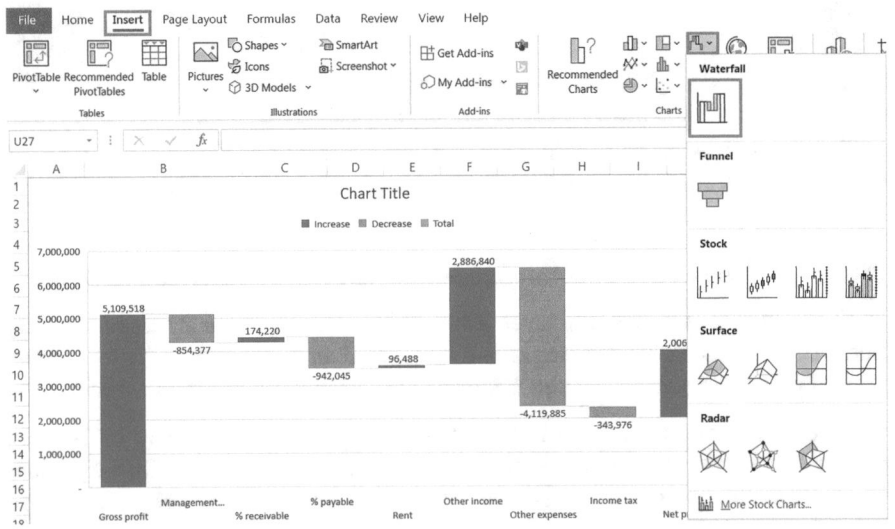

At first, it is still clear: the blue column of gross profit, management expenses are deducted from it… But instead of green growth columns, they are blue again, and at the end the net profit is hanging in the air. Yes, and in the legend there is still some incomprehensible gray "Result."

Step 2. The fact is that "Total" is the sign of the column that will be built from zero. To "land" a net profit, you need to guess first, click this segment once, and then the second time so that it lights up. Do not confuse it with a quick double click: press the left button once, release, press again. After that, right-click the context menu and select "Set as Total". We do the same for the initial column.

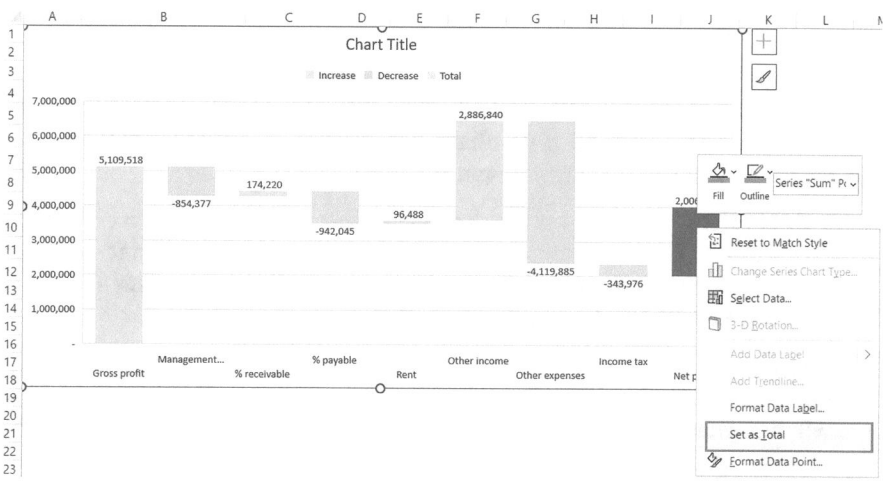

Step 3. The "bridge pillars" have turned neutral gray, but the revenues have remained blue, and I want to make them green. And then the non-obvious nuances begin again. Of course, I can select each column in turn and put the desired shade of green and red on it. But Excel provides the ability to customize the color for the entire category at once.

To do this, click the legend once, then again – on the "Fill" element and select the green color for it. This function works stably only in versions 2019+, in older versions it turns the background of the legend green.

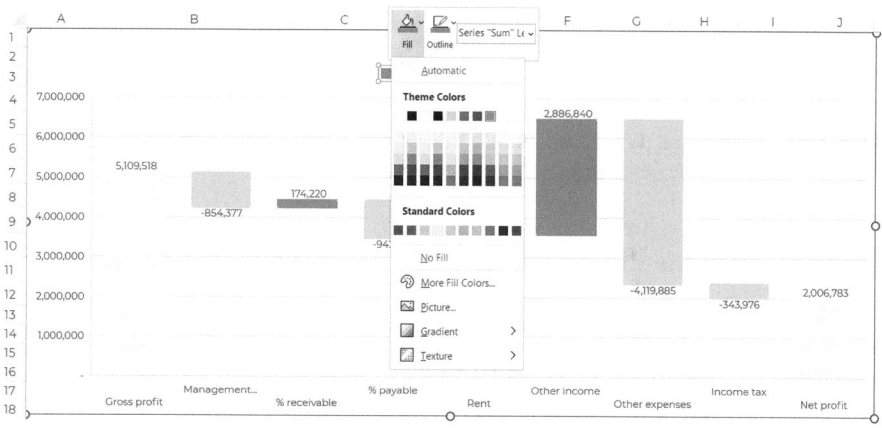

Step 4. Design the diagram according to the checklist from Part 3.4: we remove the extra Y axis, gridlines, legend. In order for the category labels to fit horizontally, you will have to reduce the font and stretch the diagram in width. If there is not enough space for the text, Excel will force it to rotate and place it at an angle. So if you have long names of deviation factors, they still won't fit, there's nothing you can do about it.

Step 5. Finishing touches. I am pleased that the gap width of the columns is 50% by default. But it is for the waterfall diagram that it is necessary to strengthen the visual metaphor and make the "steps" almost adjacent to each other. To do this, we reduce the gap to 10%.

The location of the data labels is not accidental: for the positive categories they are above the colored columns, and for the negative ones they are below it. I will not change this, I will only adjust the bit depth, remove the decimal places and slightly increase the font size. Now the chart is ready!

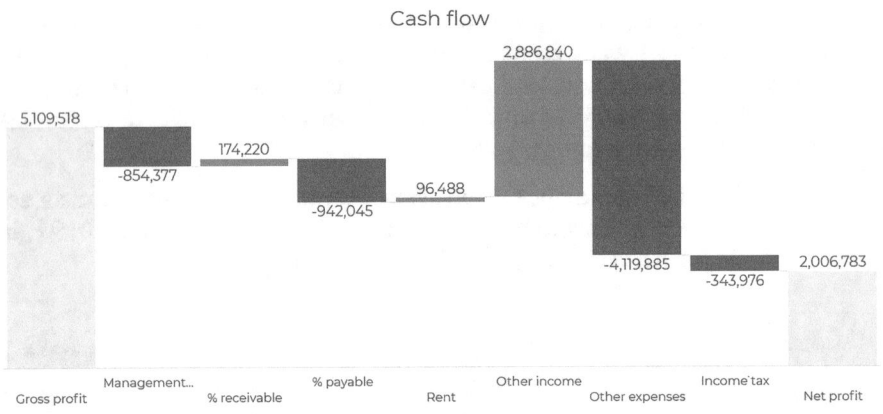

Just like a funnel, a waterfall chart is not built on pivot tables in Excel. To deceive it, again make links to neighboring cells.

How to Build a Waterfall Chart in Old Versions of Excel

If you have Excel 2013 or still 2010, you can also build the same beautiful waterfall using a "pillow." I will say even more – you will have more freedom of action. For example, you can make a horizontal waterfall based on a bar chart and accommodate long category names.

Step 1. In the table, we create the "Minus" and "Plus" columns. In the first one, we will transfer indicators with negative values from the "Sum" column (without a minus sign!), and in the second – with positive ones. In the "Sum" we leave only our initial indicator – the size of the gross profit. Based on the resulting table, we will build a diagram.

Item	Sum	Minus	Plus
Gross profit	5,109,518		
Management expenses		854,377	
% receivable			174,220
% payable		942,045	
Rent			96,488
Other income			2,886,840
Other expenses		4,119,885	
Income tax		343,976	
Net profit			

Step 2. Now we will fill the empty column "Sum" – for this we will need formulas. First, fill in the second cell of the column, under the gross profit indicator.

Enter the formula: mark the previous cell of this column ➤ subtract the value from the "Minus" in the active row ➤ add the previous cell of the "Plus" column.

	× ✓ *fx*	=C3-D4+E3		
	B	C	D	E

Item	Sum	Minus	Plus
Gross profit	5,109,518		
Management expenses	=C3-D4+E3	854,377	
% receivable			174,220
% payable		942,045	
Rent			96,488
Other income			2,886,840
Other expenses		4,119,885	
Income tax		343,976	
Net profit			

In my case, the formula will look like this: **=C3-D4+E3**. After that, it will only be necessary to stretch the formula for the entire "Sum" column. If you have calculated everything correctly, then the last two values (income tax and net profit) will be the same for you. As a result, we get this table.

Item	Sum	Minus	Plus
Gross profit	5,109,518		
Management expenses	4,255,141	854,377	
% receivable	4,255,141		174,220
% payable	3,487,316	942,045	
Rent	3,487,316		96,488
Other income	3,583,804		2,886,840
Other expenses	2,350,759	4,119,885	
Income tax	2,006,783	343,976	
Net profit	2,006,783		

Step 3. Select the table and insert a stacked bar chart. We get the base for the future "waterfall." Then we make the "sum" row transparent, but the first and last recumbent columns in this row should become totals. We select each one separately and paint it gray.

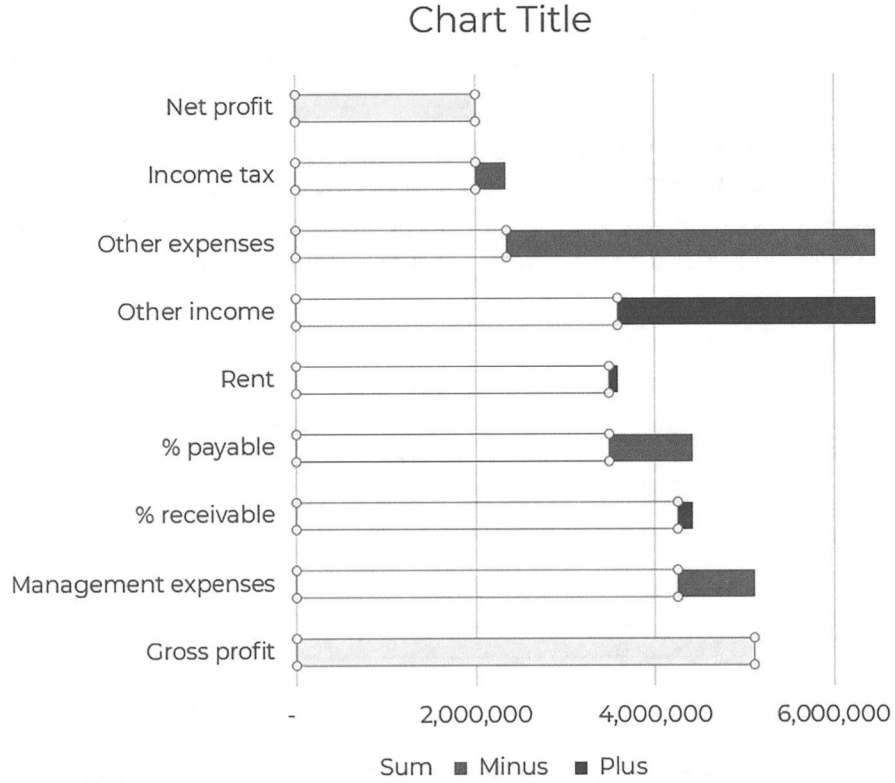

Next – setting up the design according to the checklist from Part 3.4. We get a horizontal waterfall chart, which has no analogs even in new versions of Excel. It solves the problem of long category names, because of which the inscriptions on the waterfall are cut off or rotated.

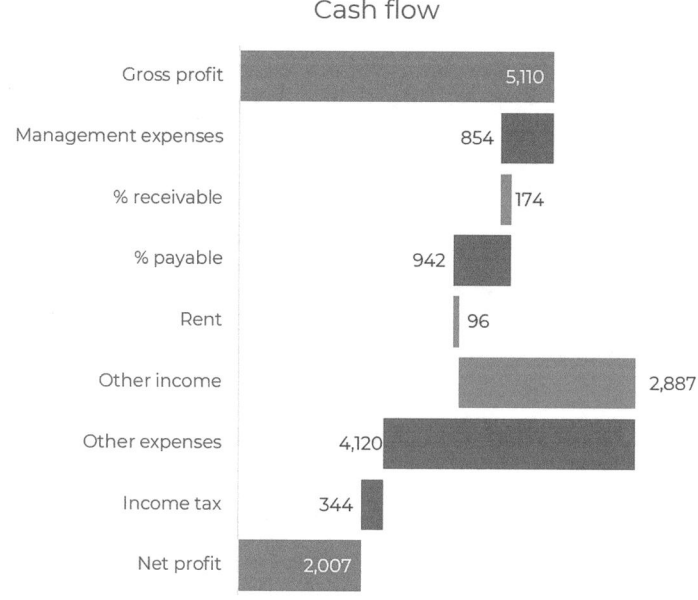

According to this principle, you can also create a vertical waterfall chart, taking as a basis a column stacked chart instead of a bar chart. If you try, it will look identical to the sample.

There are many more nuances of factor analysis visualization. For example, adding columns with subtotals or "cumulative" recalculation from the beginning of the year with filtering through pivot tables. We can devote a couple more chapters to this, but my goal is to give you a general understanding of the idea, the principle of operation, without going deep into technical details.

I also share this approach, so I stopped at two popular charts. Although there are handymen who build more complex flowcharts and connection graphs. But this requires more serious preparation of data, calculation formulas, and even macros. Of the new unusual ones in Excel, there is also a sunburst chart, but it doesn't make any sense, so I didn't consider it.

I also didn't tell you about the nuances of building scatter plots and bubble charts, because they are not used on management dashboards. On static presentation slides, you can still adjust the size and format of labels. But with dynamic data filtering, column-based shapes remain the most reliable. And then the tricks begin: then make the row transparent, then make a white font on a white background.

I think this is a useful workout for the brain. When this becomes a habit, you make the necessary visualization in ten minutes, while others spend hours trying to verbally explain their idea and coordinate layouts. This is the freedom of action and flexibility of work in the new digital world.

Summary

In new versions of Excel, advanced charts appear that display combinations of types of analysis for specific business cases. They don't work on pivot tables, but this is easily solved by putting links to neighboring cells.

Knowing the anatomy of diagrams, you can build them on the basis of ordinary columns and even get more freedom of action at the same time. The main trick is an auxiliary transparent row.

Funnel = dynamics + structure

1. Add a new column to the table and calculate the pillow values in it according to the specified formula.

2. Build a bar chart with accumulation and put the "Pillow" row first.

3. Remove the color fill from this row so that it becomes transparent.

4. Design the chart according to the checklist.

Waterfall = dynamics + rating

1. Add additional columns with negative and positive values to the table.

2. Fill in the main column with new pillow values according to the given formula.

3. Build a stacked column chart. Make the row with the pillow transparent, only return the color to the first and last columns – these are "bridge supports."

4. For deviation factors, set the colors according to the traffic light rule.

5. Design the appearance of the chart according to the checklist.

How to Improve the Data-Driven Culture in Your Organization

At face-to-face trainings, I see students' eyes light up. They become very enthusiastic – ready to redo badly prepared reporting in a progressive way, to introduce dashboards in their departments, and then in neighboring ones, so as not to correct mistakes for them anymore. And in general, to carry out a revolution up to the final reporting for shareholders, which would be based on real data.

© Alex Kolokolov 2023
A. Kolokolov, *Make Your Data Speak*,
https://doi.org/10.1007/978-1-4842-8942-6

Why Expectations Are Not Met

But a month later, my team contacts the company to find out how things are going and finds out that most of them still have everything the same. Innovations are always hindered by something: preparing the quarter reports, launching the project, waiting for the new version of the ERP system to have ready-configured analytics....

So the routine is addictive and resists change. When an inspired analyst returns to real work, he is faced with conservatism and skepticism in the spirit: "they will make him redo again anyway, why try in vain." Or there is a temptation to postpone the task, I hope that the new ERP or CRM system will have all the necessary analytical reports.

But such miracles do not happen – in 13 years of work, I have not encountered them once. At best, the new accounting system will allow you to correctly upload the actual data to Excel. You will still have to compare them with a budget yourself.

The Analytics Integration Is Not a Revolution, but an Evolution

But there are hundreds of success stories among thousands of my graduates. Someone received the support of senior management and was able to build transparent reporting. Someone, after working in a "small-town" company, got a higher position in a federal or even international corporation.

It is important to understand that the construction of an analytical system is not a revolutionary, but an evolutionary process. After installing a new program, everything will not change for the better at once: this happens gradually and requires flexibility and patience.

Especially at the transitional stage, there may be difficulties when there is no single data source, and half of it has to be collected through Excel, immediately processed and assembled in the dashboard in a garage. At such a moment, you can go back to this book, reread the necessary chapter, and remember which check mark to put in the pivot table so that everything works.

A company with a data-driven approach usually goes through three levels of analytics skills: personal, group, and corporate. This book is aimed at pumping personal skills. But business gets tangible benefits from working with data only when these skills reach the corporate level. They become part of the culture.

Personal Level As the Basis

For someone, this book opened up the world of pivot tables and visualization, and for someone else, it just refreshed Excel functions in memory. Regardless of your starting level, you should not stop there. I don't mean learning Excel so much as solving new problems.

Start by turning a regular report on your project or product into a dashboard. And if there is no such regular report, it's time to take the initiative, approve metrics and configure it.

I understand that in corporations, reporting forms come down from the top – there is little space for creativity. So that you can hone your skills, I have made training courses in an online format. You are learning advanced data processing capabilities in Excel and Power BI and are not only repeating training examples but also solving tasks yourself. You send the work to the curator for review and get recommendations: how to avoid mistakes and make the report better.

View courses

In addition to online training, the Institute of Business Intelligence has a social project – the Club of Anonymous Analysts. This is a professional community for everyone who works with reporting: from IT professionals to financial directors. We hold live meetings, webinars, and interesting visualization contests.

Another format for developing personal skills is an internship at the Institute of Business Analytics. For example, at one of them, 45 teams of interns developed dashboards for charitable foundations as graduation projects. It turned out to be interesting and useful for everyone. Interns participated in charitable activities in such an unexpected way, and foundations received valuable support in organizing reporting, both external and internal.

In general, if there is a desire, you can find a suitable format for the development of personal skills.

Group Level As a Step Forward

It's nice to hear the news when my graduates organize analytical circles, data-driven clubs, and other communities of interest. They share their experience, discuss professional issues, and invite me and other experts. This is what I call the group level of analytics skills.

To develop such skills, clients invite me to conduct corporate training. I study the company's reports and presentations and develop training tasks based on them. Employees do not work with other people's examples, but with their own data – this helps both to engage and to solve specific tasks of the company.

During a two- or three-day training, we carry out a full cycle of work: from setting a task and collecting data to calculating indicators and visualizing on a dashboard.

On the first day, we study and then practice. At the end, we invite the management to a demo session and find out what we got – just beautiful pictures or decision-making tools. This is always constructive: managers give valuable feedback, say which indicators are missing, in which sections the report needs to be filtered and what additional questions arise.

It is at this moment that the transition from the group level to the corporate level happens. Year after year, I am happy to see how a culture of decision-making based on reliable data is sprouting in such companies. No one else "covers up" the failures on the slides, everyone is ready to face the truth with the help of dashboards: who did not fulfill the sales plan, for which item there is overspending, and how to solve the problem.

How corporate training is going

Corporate Level As a Strategy

But there is another extreme. They object to me: now I will show the dashboard, and the boss will want to see the current figures every morning. Only now I will have to prepare data for him at night, because no one has canceled the main work.

This problem occurs when analytics is developed locally in one department, without synchronization with other business functions. Advanced financiers and marketers are outraged that they "have to do the work for IT specialists." Similarly, BI analysts complain that they have set up dashboards for salespeople, but they do not use them. Indeed, this should not be the case.

Why Do You Need a Data Warehouse

At the corporate level, it is important that analytical data is stored in the database, and the analyst and the manager have convenient access to it, they can use the constructor to build themselves the necessary report.

To do this, it is not necessary to train the entire team in the development of dashboards. The foundation of the corporate reporting system is laid by IT specialists. And analysts configure this system so that it works for business purposes.

It is necessary to build a data warehouse – a single database for automatic loading of data from the source accounting systems. In large companies, it is ERP, CRM, production systems. There are also marketing indicators from individual advertising cabinets and counters – the collection, cleaning, and processing of this data also needs to be organized. And I don't mention "budgets" and plans that are made up monthly in Excel, and even in a different form for each department.

All this data cannot simply be "piled up." They need to be linked together by reference cost items, product nomenclature, customer classifiers, and income and expenses should be distributed to financial responsibility centers.

This is a large amount of technical work, but it is impossible without the participation of a business analyst, a system architect. On the one hand, he understands the logic of calculating indicators, and on the other, he knows how to shift this to the structure of the database and data flows.

How I Work As an Interpreter Between IT and Business

The Institute of Business Intelligence also performs work where it is important to solve technical problems and make data work for its clients' business.

In order not to lose my qualifications, I continue to lead key projects personally. I immerse myself in the company's strategy, discuss tasks with directors and owners in a language they understand. And then I translate all the requirements into the language of the developers. In the same way, my analysts interact with each owner of the business process and data source.

To be honest: as a result of this work, the client does not always get non-obvious business insights based on statistics. Most often, we uncover inefficient areas that remain unnoticeable when the company's turnover is already measured in billions.

Somewhere it turns out that the supplier was overpaid for raw materials at an inflated price. Somewhere, finally, we saw a negative margin due to too much discount to the client. It's not necessarily someone's malicious embezzlement:

most often the reason is that neighboring departments only learned about the overall results once a quarter and did not understand how their work affects the rest.

We do large turnkey projects for some customers; from the integration of accounting systems to the development of dashboards and training. Others have their own IT team, and we work out business requirements, design layouts, and technical tasks. At the same time, we carry out author's supervision. And someone applies for one-time consultations when it is necessary to solve a complex non-standard task.

I want my work to bring maximum benefit to clients, whether it's three months of work or two hours of consultations. All the services and products of the Institute of Business Intelligence are aimed at this. Read on the website how we can be useful to companies and their managers.

BI implementation and dashboards for business

Your Data Must Speak the Language of Business

There is a phrase on the website of the Institute of Business Intelligence: "Saving the world from useless reports." And I really consider this my mission, no matter how pathetic it may sound. I don't just believe that data-driven management helps businesses grow. I know that.

So, the question is how to present this data clearly. I hope that the book has helped you solve this difficult task.

For analysts – To learn how to look at data from the top down, from a business perspective, packaging your reports beautifully and expensively, like a complete information product. And don't waste a lot of time on it.

Middle managers – To convey their ideas convincingly present them with the help of modern interactive dashboards, involving their colleagues and partners.

For directors and top managers – To raise the culture of working with data to a new level. So that you don't have to wait weeks for reports and then sort them out on your own.

And everyone – To let your data speak. And it will help you make the right decisions in business.

Index

I

© Alex Kolokolov 2023
A. Kolokolov, *Make Your Data Speak*,
https://doi.org/10.1007/978-1-4842-8942-6

Printed in the United States
by Baker & Taylor Publisher Services